D1483840

A Simpler Life

EXPERTISE

**CULTURES AND
TECHNOLOGIES
OF KNOWLEDGE**

EDITED BY DOMINIC BOYER

A list of titles in this series is available at cornellpress.cornell.edu.

A Simpler Life

Synthetic Biological Experiments

Talia Dan-Cohen

Cornell University Press
Ithaca and London

Copyright © 2021 by Cornell University

All rights reserved. Except for brief quotations in a review, this book, or
parts thereof, must not be reproduced in any form without permission in
writing from the publisher. For information, address Cornell University
Press, Sage House, 512 East State Street, Ithaca, New York 14850. Visit
our website at cornellpress.cornell.edu.

First published 2021 by Cornell University Press

Library of Congress Cataloging-in-Publication Data

Names: Dan-Cohen, Talia, 1982– author.
Title: A simpler life : synthetic biological experiments / Talia Dan-Cohen.
Description: Ithaca [New York] : Cornell University Press, 2021. |
    Series: Expertise | Includes bibliographical references and index.
Identifiers: LCCN 2020032521 (print) | LCCN 2020032522 (ebook) |
    ISBN 9781501753442 (hardcover) | ISBN 9781501754333 (paperback) |
    ISBN 9781501753459 (pdf) | ISBN 9781501753466 (epub)
Subjects: LCSH: Synthetic biology—Methodology. | Synthetic biology—
    Research—Philosophy. | Technology—Social aspects.
Classification: LCC TA164 D36 2021 (print) | LCC TA164 (ebook) |
    DDC 660.6072—dc23
LC record available at https://lccn.loc.gov/2020032521
LC ebook record available at https://lccn.loc.gov/2020032522

Contents

ACKNOWLEDGMENTS

My greatest debts are owed to those who permitted me to study their labs. Michael Hecht and Ron Weiss shared their time, space, and thoughts with me, exhibiting both generosity and curiosity. Many of their students and postdocs followed suit. Since I have anonymized those who were more vulnerably situated, I regret that I can acknowledge their generosity and grace only collectively. It goes without saying that this project would never have gotten off the ground, much less sustained flight for so many years, without this group of wonderful interlocutors.

Princeton University proved both an interesting site and a convivial home in which to give this project its initial form. Abdellah Hammoudi's intellectual guidance was formative in ways that both permeate and exceed this book. Jim Boon, Rena Lederman, and Larry Rosen read, listened, taught, and inspired. Mentoring at Princeton was wonderfully collective. João Biehl, John Borneman, Lisa Davis, Carol Greenhouse, Alan Mann, Carolyn Rouse, and the late Isabelle Clark-Deces helped make Princeton's anthropology department the vibrant place that it was during my years there. Completion of a very early version of this work was supported by Princeton University's Woodrow Wilson Society.

I am exceedingly grateful to colleagues at Washington University in St. Louis for supporting and engaging with this work at various stages. Among them, John Bowen has proved an invaluable mentor and interlocutor. Beyond the anthropology department, colleagues from across campus have made life at the university much more interesting and fulfilling. A special thank-you to participants in the Philosophy of Science Reading Group, Ethnographic Theory Workshop, and Social Studies of Institutions Group for offering consistent intellectual enrichment and for engaging with several portions of this work.

This book has benefited from conversations with many colleagues, friends, and mentors over many years, including Mark Alford, Gaymon Bennett, Peter Benson, Venus Bivar, Leo Coleman, Carl Craver, Tili Boon Cuillé, Monica Eppinger, Bret Gustafson, Nathan Ha, Ronnie Halevy, Janet Hine, Helen Human, Jean Hunleth, Peter Kurie, Rebecca Lester, Richard Martin, Stephanie McClure, Paige McGinley, Claire Nicholas, Bruce O'Neill, Shanti Parikh, Anthony Petro, E. A. Quinn, Tobias Rees, Annelise Riles, Mark Robinson, Elizabeth Schechter, Lihong Shi, Larry Snyder, Priscilla Song, Anthony Stavrianakis, Glenn Stone, Kedron Thomas, Erica Weiss, Ian Whitmarsh, Mo Lin Yee, and Carol Zanca. I owe a tremendous debt of gratitude to Paul Rabinow, without whom this book (not to mention the trajectory that begot it) would be unimaginable.

At Cornell University Press, Dominic Boyer and Jim Lance helped make the publishing experience a truly pleasant one.

I have been immensely lucky on the family front. My deepest thanks to Paulo, Mina, Leo, Hana, Meir, and Ishai for all of the love and humor that sustained me throughout.

An earlier version of chapter 2 appeared as "Ignoring Complexity: Epistemic Wagers and Knowledge Practices among Synthetic Biologists," *Science, Technology, & Human Values* 41, no. 5 (2016): 899–921, doi:10.1177/0162243916650976.

A Simpler Life

INTRODUCTION

In 2004, J. Craig Venter and Daniel Cohen, two famed geneticists, published a review that began with a declaration: "If the 20th century was the century of physics, the 21st century will be the century of biology."[1] This bombastic opener set the tone for a piece that celebrated the promises of the postgenomic age in the quasi-religious language of destiny. The authors wrote, "In a very real sense . . . man will reach the final frontier of his own fate when, in the Age of the Genome, he possesses the blueprint to redesign his own species."[2] A self-fashioned genomics maverick, Venter had headed the private effort to sequence the human genome at Celera Genomics, competing with the publicly funded Human Genome Project. The sequence was, we recall, likened to the Holy Grail. It was expected to answer a lot of questions about living things, to explain differences and provide drug targets, to allow a glimpse into our own individual biological futures and our collective evolutionary histories. The buildup was not quite matched by the pace of discovery in the wake of the sequencing project. Many biotech companies formed and dissolved, their research agendas and business plans unwieldy in the face of anxious investors, recalcitrant genomes, and mountains of sequence information.

Sequencing genomes was what had gained Venter his considerable fame, yet it is not the primary endeavor to which he has hitched the destiny of mankind. In 2004, the field of "synthetic biology"—an umbrella term for attempts to design and assemble new life-forms—was just gaining speed. Venter was himself mounting an impressive effort to make headway in this new domain through research at the J. Craig Venter Institute. The activities linked together under the heading "synthetic biology" have always been heterogeneous, but one underlying theme uniting many of them is a bottom-up approach to the design and construction of synthetic organisms. Elaborate a biological toolkit. Forget, black-box, or defer the complexity. Build a better, simpler life.

Though synthetic biology got its start in the early 2000s, it wasn't until 2010 that Venter and his institute managed to reach a major milestone in this new pursuit. In June of that year, Venter's team announced the successful synthesis of a very simple life-form: scientists at the institute had stitched together a synthetic genome almost identical to that of the bacterium *Mycoplasma genitalium* and transplanted it into a host cell that expressed the donor DNA. In countless newspapers and journals, the construction of this being was tallied as a win for the emergent field. More a proof-of-concept than a technological breakthrough, the bacterial hybrid nonetheless provided an occasion for scientists and news agencies to hail the arrival of designer synthetic organisms that will someday eat environmental contaminants, produce fuel, provide shelter, and even search for and destroy diseased cells in our bodies.[3]

Many of the articles announcing the successful synthesis also ran an accompanying photo of a pair of these new life-forms, slightly asymmetrically situated on a fleshy orange background, resembling two eyes peering back at the reader as though through bright-blue irises.

What to make of this reciprocal gaze? In a piece on the contemporary biosciences, written in the late 1990s as the Human Genome Project was gaining speed, Paul Rabinow noted that the unanticipated overlap between the human genome and gene sequences present in other animals, shuffled and recombined, lent a certain prescience to Rimbaud's cryptic prophecy that "the man of the future will be filled with animals."[4] Yet, just as the human seems to disaggregate into an evolutionary history of recombinable bits of animality, organisms themselves increasingly become artifacts of human design. We

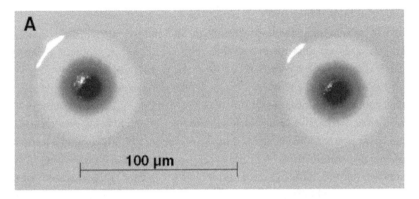

From Daniel G. Gibson et al., "Creation of a Bacterial Cell Controlled by a
Chemically Synthesized Genome," *Science* 329, no. 5987 (2010): 52–56.
Reprinted with permission from AAAS.

might want to ask, then, in response to Rimbaud's prophecy, with what exactly will these animals themselves be filled?

From certain angles, this question seems relatively new. Humans have long attempted to situate themselves in the order of things by reference to the mirror of nature, a nature that also peeks out from within. So what happens when biotechnology remakes the natural world? The ensnaring, expressionless stare of these two eyes captures something of the seeming contemporary existential predicament of culture seen through the mirror of culture (in cell culture, no less). In his writings on the frontiers and borders of the biosciences where the limits of "life" are encountered, Stefan Helmreich argues that, "in the age of biotechnology, genomics, cloning, genetically modified food, and reproductive technology, . . . when human enterprise rescripts and reengineers biotic material, a founding function for nature is not so easily discernible. Culture and nature no longer stand in relation of figure to ground. Life forms cannot unproblematically anchor forms of life."[5]

Yet, from another angle, the predicament is decidedly not new. Critical observers have argued that nature has always furnished culture with exceedingly shaky—because malleable—ground.[6] Nature is in important ways already made, not only through collective ways of ordering and classifying, of drawing out and imposing norms and relations, but also in the specificity of expert voices; of who gets to know it, and speak for it, and how; of what practices get to count, what concepts to stick; of what observations to hold; and of what projects it can be recruited into and at what costs.

In this book, I take synthetic biology to be the latest permutation in a history of mutual incursions between nature and culture, and a contested, heterogeneous, and unstable one at that. The "synthetic" in synthetic biology can be thought of not only in reference to the synthesis of new life-forms but also in relation to the tenuous gluing together of disparate sorts of stuff (e.g., concepts, rationalities, expert practices, institutions, materials, techniques, aesthetics, forms, and values) that accompanies efforts to bring new techno-epistemic projects into existence. This book is about a highly curated selection of this disparate stuff, seen at a moment of extraordinary uncertainty and viewed at ethnographic close range. The fieldwork for the book took place in two synthetic biology labs at Princeton University over the course of three years, between 2008 and 2011, years in which, as I show, the tackiness of the glue holding together a hastily synthesized field was still very much in question. Today, that glue has still not quite dried.

Making Life Is Easy

A few years after Venter and Cohen made their grandiose prophecy about the century of biology, I sat in the audience while the head of one of the Princeton labs in which I conducted my fieldwork, Michael Hecht, a chemist and synthetic biologist, gave a talk about his lab's research that ended with a decidedly more measured—if ironic—assessment.[7] Michael followed the substantive portion of his talk with a slide containing an image he had silk-screened onto a T-shirt for a graduate student as a parting gift. The T-shirt bears an image of a petri dish smattered with bacterial cell colonies and, underneath the dish, the inscription, "Making life is easy. It's making a living that's hard."

A few interpretations of Michael's slogan set the scene. Life is easy to make today.[8] In the aftermath of the Human Genome Project, technological advances have continued to lower the cost of sequencing DNA at an incredible pace. Standardization and marketization of the technical means necessary to synthesize DNA and tinker with genetic codes have increased access to these technological goods and simplified their use, inviting an assortment of practitioners to join in the project of radically modifying life-forms.[9]

Meanwhile, genetic data has largely thwarted some of the parsimonious models guiding expectations about the relationship between DNA sequences and the living things we see around us. In 2013, the Nobel Prize–winning biologist Sydney Brenner lamented, "We are drowning in a sea of data, and thirsting for knowledge."[10] The gulf between information, which was amassed in abundance by machines that churned it out at an ever-increasing pace, and understanding or knowledge was becoming increasingly apparent. Into that gulf poured new kinds of biological endeavors—proteomics, epigenetics, metabolomics—all concessions to the fact that the genome emerged at the beginning of the twenty-first century as a greater mystery than it had seemed in decades prior.

Roughly twenty years before Brenner summed up the postgenomic situation in terms of drowning and thirsting, in the essay that cited Rimbaud, Rabinow had noted the specifically modern rationality driving the Human Genome Project, then at its inception. "The object to be known—the genome—will be known," he observed, "in such a way as to be *changed*. This dimension is thoroughly modern; one could even say that it instantiates the definition of modern rationality. Representing and intervening, knowledge and power, understanding and reform, are built in, from the start, as simultaneous goals and means."[11]

In some ways, synthetic biology captures this modern rationality well. Yet Brenner's "thirst" suggests that in the postgenomic era, the relationship between representing and intervening, between knowledge and power, is proving more difficult to pin down. As Matthias Gross has observed, "The modern idea of science as a means for turning uncertainty into certainty instead has more often led to more knowledge about what is unknown and perhaps cannot be known."[12] Where technoscience flourishes, ignorance proliferates.

What is intervention in the absence of knowledge? The historian and anthropologist Sophia Roosth has argued that synthetic biologists are engaged in a rhetorical loop in which they make new biological things in order to understand the things they have made.[13] Roosth's summary of the relationship between knowledge and intervention in synthetic biology nicely captures an aspirational register. It also moves us along quickly past the opacity that currently plagues such things, and that makes them both daunting and fecund. For if Roosth's rhetorical loop is read as a process that must necessarily unfold over time, then it contains the intriguing suggestion that, in the meantime, synthetic biologists don't understand what they make; new biological things are, at least partly, shots in the dark. Moreover, such shots in the dark must do more than satisfy the popular desire to master the momentary spark of life, to legislate the line between the animate and the inanimate. They must exhibit the ability to live in and interact with an environment predictably and remain constant over time, requirements that increase the dimensions of the ignorance that surrounds them precipitously. The gap, therefore, between these new biological things and either knowledge or technological applications is substantial, and the ways of bridging it experimental, and often quite local. In this book, rather than try to diagnose and affix the relationship between knowledge and intervention in synthetic biology generally, I attend to various "techno-epistemic" practices and forms of reason with which practitioners attempt to maneuver through vast swaths of ignorance by leveraging different relations between knowledge and intervention. Such an approach seems well suited to an unstable technoscience, whose viability is still something of a question, whose normative coordinates are still unclear, and in which, as I was told on numerous occasions, "almost everything has been harder than anticipated."

Michael's T-shirt slogan is also a salutary reminder to separate hubris and rhetoric from the predominant conditions of research in synthetic biology. Synthetic biology is often described as industry-oriented. Though there is a veritable smorgasbord of start-ups and companies devoted to synthetic biology, the infrastructure of the field is in many ways still cobbled together and heavily academic. Some synthetic biologists are trained engineers. Some are scientists. Some synthetic biologists are homed in industry. Many populate assorted engineering and natural science departments and navigate an existing ecology of knowledge practices, norms, and values, as well as the usual gauntlet of academic and career credit and attendant academic hierarchies. This is partly because, in recent history, technological innovation has steadfastly relied on research universities to supply the forgiving home for exploratory research that the market often will not. Many of the most visible practitioners speak from the academic pulpit (though they often also dabble in industry), which they frame as an important center of gravity for producing standards, textbooks, and stability. In 2013, in a piece that ran in *The Chronicle of Higher Education*, tellingly titled, "Synthetic Biology Comes Down to Earth," a leading synthetic biologist lamented the rush to industry among an early crop of graduate students in the budding field, registering his sense that it would be better to keep more talent in the university while the field is still young and unformed.[14] Note, in the metaphoric landscape of the article, universities are decidedly terrestrial, the place synthetic biology comes back *down to*. With its more plodding temporalities and lumbering gait, the academic world is thus a unique protagonist in many stories we might tell about synthetic biology, the ones told in the pages that follow among them. It is in the university that, among other things, I set a variant of a question Max Weber posed of science early in the last century: What are the conditions, external and internal, of technoscience as a contemporary academic vocation?[15] In exploring this question, I highlight and analyze some of the specific rationalities underlying research endeavors, as well as the institutional dynamics and structures that enable, constrain, and discipline the technoscientific career.

If synthetic biology is frequently framed as industry-oriented, it is also often described as fabrication-focused, wherein fabrication replaces experimentation as a mode of generating and engaging with novel biological systems. Yet the activity that many who view themselves as pursuing synthetic biology are engaged in is experimentation. This observation merits some emphasis. While many synthetic biologists publicly fantasize about a future in which their field will be primarily concerned with the design of biotechnological systems, and in which factories will take over much of the construction work currently done in labs, synthetic biology remains an experiment-driven field. As a number of observers have noted, the day-to-day dimensions of synthetic biological research can be remarkably similar to what one would see in a molecular biology lab.[16] One technique social science observers have employed to deal with the overlap involves omitting points of similarity and positing the remainder as what is new and interesting about the field. In contrast, the approach I take here is to insist that an ethnographic examination of what synthetic biologists do at and near the lab bench provides a valuable perspective despite, or perhaps even because of, the way that it threatens the coherence of synthetic biology as a bounded object. Many of the discussions contained in this book therefore properly relate to the sociology and philosophy of experimentation. My engagement with these areas and literatures is intended to make the point that synthetic biological practices intersect long-standing questions about experimental practices generally, rather than rendering them moot.

Up and To the Side

If Michael's T-shirt slogan captures something of the substantive ground to be covered, it also inspires a certain ethnographic positioning and tone. In his book *Never Pure*, Steven Shapin describes his own scholarship over the past few decades as aimed at, among other things, "lowering the tone."[17] Shapin uses the phrase, roughly, to place his own efforts among a wide range of studies in the history of science that have sought to reevaluate the almost deistic virtue of Science, to lower the capital S and focus on the heterogeneous, and very situated, set of practices that constitute sciences. New times, new deities. Like Science—of the capital S variety—synthetic biology is often addressed in a high tone rooted less in notions of virtue than in hopes for technological salvation (or worries about collective doom). And, as with the sciences, the high tone circulates both in the cultural ether and among a reflexive and vocal cast of leading practitioners.

The reflexive component seems particularly intense among synthetic biologists, who, in penning their own review essays assessing and reassessing synthetic biology's progress and declaiming the history of their field, also seek to stabilize a network. If Niklas Luhmann gave social scientists the role of second-order observers who observe observers observing, attempting this positioning with synthetic biologists, if one internalizes some leading practitioners' more totalizing moves, yields the strange sense of observing observers who are the observers of their own observing, steadfastly folding descriptions into the object—synthetic biology—they seek to assemble.[18]

The self-conscious process through which synthetic biology is composed as a thing-in-the-world is reminiscent of an observation made by Marilyn Strathern in her work on university audit practices. Strathern identifies a contemporary Euro-American cultural form so "habituated to description as an artefact of visibility," that practitioners themselves "produce second-order accounts of their activities, and produce these as indeed accounts of society and culture, of accounts of how they were organised and thus as accounts of *their own organization*."[19] "Anything an observer added," continues Strathern, "would then become a new second order activity, another way of making it visible, through redescription, until, that is, the new description became likewise absorbed into the organisation's knowledge of itself."[20] Strathern's observation, which resonates with insights from actor-network theory, directs our attention toward imbrications between description and construction.

When writing about a hot and hyped emergent technoscience that seeks stabilization through description, tone-lowering isn't merely a means for providing a reality check. It does important epistemological work. In adopting Michael's slogan as a framing device for this book, I mean to operationalize perspectives that cut across and against the descriptive vortex that surrounds synthetic biology. I have attempted to enter synthetic biology at an angle. Toward this end, I have made use of my ethnographic locale: Princeton University, a Euro-American research institution where the "Euro" is also aspirational; the westward-cast gaze is built into the university's Oxbridge-style gothic architecture. In my study design, I had initially left open the possibility of switching locales to pursue a more conventionally global view of synthetic biology in what would be termed, among anthropologists, a "multisited" project. Multisited methods, in which the anthropologist is set free to follow an object or formation, are now common for anthropological studies of science and technology, responding to a concern that the traditional methods of the discipline, specifically the dependence on the single, bounded locale, are out of step with both a changing world and with the forms of representation available for anthropologists. This view, and the anxious pathos attached to it, is best summed up by Michael Fischer's wonderfully evocative and oft-cited line, decrying the predicament shared by diverse contemporary fields and practitioners, that "the traditional concepts and ways of doing things no longer work, that life is outrunning the pedagogies in which we were trained."[21] This outrunning, in turn, Fischer suggests, is clearest at the frontiers of technoscience, where new biotechnologies, data banks, network infrastructures, and ecological changes reshuffle relations among capital, institutions, and social orders. In Stefan Helmreich's writings on marine biologists, Fischer's observation is deployed as a pun, wherein it is the concept of "life" that does the outrunning, indexing the broader unmooring of culture from nature mentioned earlier in this introduction.[22]

George Marcus has written a series of essays on "multisited" field imaginaries in anthropology that motivate such methods in interesting ways, including as a push against what he terms "resistance and accommodation studies" that pit a locale against some macro-frame like "global capitalism."[23] Marcus advocates for an ethnographic experiment that attempts to "blur the macro-micro dichotomy in trying to represent both place and system in multiple perspectives."[24] Yet, as is wont to happen to well-thought-out methodological innovations when they become routinized, their utility sometimes comes unmoored from the textual and representational settings that had made them cogent interventions. In practice, "multisited" ethnography often responds to the perceived spatial limitations of "traditional" ethnography when brought to bear on objects figured in such a way as to exceed our notion of a site.

As Matei Candea has astutely observed, there is a "problematic reconfiguration of holism implicit (and sometimes explicit) in the multi-sited research sensibility—a suggestion that bursting out of our fieldsites will enable us to provide an account of a totality 'out there.'"[25] When that totality is an emergent discipline like synthetic biology, the problem with smuggled holism is that it conjures, reifies, and concretizes what are still ephemeral links, connections, and formations, ones that may in truth still fail, but that many actors have a vested interest in making concrete. To put this in actor-network theory terms again, the anthropologist gets recruited as an ally in the project of stabilizing the network.

Feeling the seductive pull of the network, I chose instead to double down on my particular locale. Princeton provides a peculiar perspective; partial, like every perspective—multi or not—yet a perspective nonetheless. Princeton University was not a hotbed of investment in synthetic biology, nor was it particularly strong in the more general area of biological engineering during the years of my fieldwork. The university had attempted to hire a very prominent synthetic biologist a few years before I started hanging out in the labs, but the hire had fallen through, at the same time that other synthetic biology practitioners or dabblers had left. The university was not following the international trend toward the speedy creation of bioengineering programs, schools, and degrees. Both of the principal investigators at the center of this book took part in efforts to launch a bioengineering program driven by a collection of interested faculty. On the administration's side, uptake was, I was told, frustratingly slow.

In their own collaborative project with synthetic biologists, Andrew Balmer, Katie Bulpin, and Susan Molyneux-Hodgson locate the synthetic biologists they study in the budding field's "periphery," a spatial and hierarchical metaphor I find useful, and which captures salient features of this book's ethnographic setting.[26] Among their project's features, their distinction between "center" and "periphery" captures how practitioners themselves talk about the relative positioning of different research sites. It also reflects differential "authority publicly to define what synthetic biology is."[27] "Center" and "periphery" thus organize a self-reinforcing symbolic geography.[28]

Princeton's particular research affordances for synthetic biology are consistent with the status of an *elite periphery*, on the map and yet off it, a site of privilege, prestige, and relative isolation. In this book, I am certainly "studying up," a term Laura Nader coined as a catchall for studies of those who, in contrast to the traditional subjects of anthropology, wield some power.[29] But my positioning also makes clear that not all "ups" are the same. My study is focused somewhere in the middle-up and a little off to the side, where many labs are situated. It is precisely the relative isolation of the elite periphery that proves useful for detaching from the inadvertent labor of stabilizing a network and its attendant hype sphere, and for retrenching description in daily life, in what people do, and how they reason about what they are doing. The elite periphery is, then, a critical positioning both in relation to multisited field imaginaries and in relation to the tropes of speed and connectedness that attach themselves to hype technoscience.

The labs whose stories are chronicled in this book serve the purpose of what Candea terms an "arbitrary location": a "methodological instrument for deferring closure and challenging totality."[30] Candea's model system for thinking through "arbitrary locations" is the village, a classical field site whose boundedness was challenged through a series of critiques in the 1990s. Candea therefore takes up the village with "calculated irony."[31] His message is clear: the limitlessness and seamlessness of our networked world are aesthetic tropes in their own right, often as superficial as the bounded field site.

The village and the laboratory share certain features. For one, early laboratory ethnographies, especially those by science studies scholars and sociologists, often drew quite specifically on the rhetorical and epistemological repertoire of anthropological writings about native villages and tribes.[32] This somewhat hokey analogy achieved rhetorical estrangement from the supposedly familiar activities of scientists, rendering them exotic at and after a moment when exoticism itself was coming under fire in ways that reduced it to a rhetorical construction. Irony was therefore built in from the start.

By now, lab ethnographies, too, have gained "village" status. They are ubiquitous, yet quaint. And the boundedness of the lab, too, has been challenged. Yet in disentangling my study from the breathy futurism that surrounds synthetic biology, the strategic quaintness of the lab study situated in the elite periphery has proven generative. Of course, depending on the units counted, my "two labs" is multi, too. Arbitrary locations, however, are not about reducing the number of sites but about dislodging an imaginary that sneaks holism back into anthropology, this time tethered to the network tropes of a fast-paced, complex, globalizing, interconnected world that are both constitutive of that world and derived from it.[33]

I have attempted to enter synthetic biology at an angle in some other ways as well, ones improvised through a number of means. Chapter 2, for example, approaches the relationship between knowledge and intervention in synthetic biology by considering the uses of ignorance among practitioners; chapter 3 focuses on ambiguous experimental results; chapter 4 unpacks an unsuccessful attempt to publish a key paper; chapter 5 follows a lab through the seemingly mundane process of moving from one building to another. Perusing the chapters, one finds that many of the events and episodes highlighted involve day-to-day hiccups and impasses on the way to synthetic biology. The overall result is a set of descriptions and analyses that are a little hard to assimilate into the work of objectification or into triumph or doom narratives, and that leaves open the question of the field's coherence as an object, an openness well suited to a study of preliminary experimental steps in an uncertain field.[34]

What Is Synthetic Biology?

Still, we will require a starting point. We may recruit an extraordinarily broad definition of synthetic biology with some ease: designing and constructing novel life-forms.[35] Efforts to pin down synthetic biology with more specificity splinter in numerous ways.[36] Media approaches to defining the field have tended to rely on a few relatively successful exemplars from the early and middle 2000s that suture together specific approaches, personas, and institutions. To sample some of these exemplars, we peruse the pages of volume 403, number 6767, of the journal *Nature*, published in January 2000. The news section reports on pressing problems of the day. Celera Genomics' licensing terms for genetic information raise fears of monopoly; a US National Research Council panel concludes that global warming is "undoubtedly real"; Democratic congressman Henry Waxman voices dismay at the underreporting of adverse effects in NIH-funded gene therapy trials in humans.

Nestled deep in the pages of this issue are two research articles with the titles "A Synthetic Oscillatory Network of Transcriptional Regulators"[37] and "Construction of a Genetic Toggle Switch in *Escherichia coli*."[38] Hardly noteworthy to the uninitiated, the publishing of these articles side by side is often pinpointed as a crucial moment for what has been termed the engineering or parts-based approach to synthetic biology. The first of these papers, written

by Michael Elowitz and Stanislas Leibler, describes the design and implementation of a genetic oscillator, a network of genes strung together to cause
E. coli cells to periodically fluoresce. The second, written by Timothy Gardner, Charles Cantor, and James Collins, chronicles the successful construction of a synthetic genetic "switch," also in *E. coli*. The articles talk about
"network architecture" and "design principles," terms mostly foreign to the
study and manipulation of living beings. Peeling back the jargon, the research
detailed in both of these papers involves picking a simple, well-characterized
abstract network type from electrical engineering (e.g., oscillator, switch, and
so on), genetically engineering cells to send and receive chemical signals,
modeling their interactions using mathematical and computational tools im-
ported from engineering, and then testing the construct in a petri dish.[39]

The research that begot these publications and others like them was undertaken in the mid- to late 1990s, largely in Boston. The most program-
matic work, so to speak, aimed specifically at founding an engineering field
with life as the construction material came out of MIT, where engineers
groomed an analogy between biological systems and electronic ones, inspir-
ing efforts to rewire gene regulation in ways that mimicked the logical
functions of common electronic circuit components. The regulatory networks described in the *Nature* papers were thus viewed as both first steps
and proofs-of-concept for an attempt to make living substance amenable to
rational design by building predictable and standardized biological "parts"
and "devices" that can be informationally and materially circulated among
labs or, more ambitiously, someday, housed and assembled in some faraway
manufacturing plant.

The epistemic well from which the engineering approach to synthetic biology draws was dug largely in the 1960s.[40] In 1961, Francois Jacob and
Jacques Monod published a groundbreaking paper in which they posited
regulatory circuits that tune how a cell responds to its environment.[41] At the
time, the central dogma, a hypothesis that holds that each gene makes one
protein, was enjoying tremendous success in explaining the function of
the material of heredity. Jacob and Monod's regulatory circuits showed that
only some genes could be explained in this way. Other segments of the genetic code were tasked not with specifying the structures of proteins, but
rather with determining the rate at which these structural genes were tran-
scribed.[42] Over the following years, as scientists interrogated the details of

transcriptional regulation in bacteria, the possibility of engineering novel regulatory networks became increasingly concrete.

The development of molecular cloning and polymerase chain reaction (PCR) in the 1970s and 1980s enabled the practice of targeted genetic manipulation. Not until the genetic data deluge of the 1990s did scientists begin cataloging lists of seemingly modular genetic components that could be shuffled around and strung together to forward-engineer regulatory networks. And so, by the late 1990s, a few scientists and engineers had begun building simple biological systems out of these components. Origin stories for this approach abound among a vocal central cast, suggesting that the analogy between electronic circuits and biological ones was ready for mining when the technological conditions would permit it. Thomas Kuhn famously used the phenomenon of simultaneous discovery to argue against heroic accounts that frame discoveries as genius leaps made by great minds. Kuhn's take on the matter, which has dominated historical accounts since the 1960s, essentially fleshes out an adage: "The time must be ripe."[43]

Another approach to synthetic biology retains much of the talk of standardization, hardware, and software but focuses on reengineering organisms to produce chemical compounds. This type of synthetic biology is often likened to metabolic engineering (and behind closed doors it is sometimes in fact equated with metabolic engineering as a way of contesting the proper limits of synthetic biology). One key figure in this approach is a UC Berkeley engineer named Jay Keasling—the key early achievement the synthetic production of an antimalarial compound. In 2000, Keasling was searching for a target molecule that could be manufactured by rewiring the metabolic pathways of cells when a graduate student showed him a recent paper on amorphadiene synthase. Amorphadiene is a precursor to artemisinin, an antimalarial normally derived from sweet wormwood, a plant native to Eurasia that is difficult to grow and financially out of reach for those who need it most. Combining genes from three different organisms and inserting them into *E. coli*, Keasling and his team's first major success culminated in a publication in *Nature Biotechnology* in 2003 that helped them secure a large grant from the Bill and Melinda Gates Foundation. Ambitiously pursuing industrial-scale production of artemisinin, Keasling and some collaborators started a new not-for-profit company, Amyris Biotechnologies, that would focus on streamlining production and increasing efficiency. Joining forces with One World Health, a San Francisco–based nonprofit specializing in production of generic drugs, Amyris and Keasling entered into an agreement with Sanofi-Aventis, a Paris-based pharmaceutical company that would make the drug. Bringing the drug to market proved a lengthy and somewhat tortuous process, owing largely to the vulnerability of both patients and farmers, but the synthetic drug was finally made available in 2014.[44] In the meantime, Keasling's research efforts have migrated to synthesizing biofuels, while his managerial acumen has been put to work in several sites. Since 2006, he has been the director of Synberc, an NSF-funded synthetic biology research center headquartered at UC Berkeley; an associate director at Lawrence Berkeley National Laboratory; and CEO of Joint BioEnergy Institute, mixing research and institution-building, like many of the founding practitioners.[45]

And then there was Venter. Venter's legacy is still in the making. Perhaps the most prominent figure in the now more than decade-old race to sequence the human genome, Venter consistently garners attention for his pursuits, and some amount of trepidation. He is the second most frequently mentioned scientist in a recent recombinant DNA textbook.[46] His ventures have jump-started or accelerated regulatory debates, partly because of his chomping-at-the-bit patent approach and partly because his personal ambition and style does not fall in line with the requisite caution we like to see from scientists, much less scientist-entrepreneurs. The *New York Times* dubbed Venter "the Richard Branson of biology," conjuring an image of millionaire adventurer CEOs seeking profit and changing the world in one stride.[47] *Forbes* magazine, meanwhile, called Venter "the Bono of Genetics."[48] He is, no doubt, a charismatic figure, a force of sorts in the genomics world and a popularly recognizable presence in a biological endeavor that has attempted for a century to erase the traces of the hands that have shaped it. Stephen Hilgartner summarizes the sentiment aptly when he writes of Venter, "With his countless admirers and critics, he is probably the closest thing biology has to a Hollywood celebrity."[49] Taken together, these analogies tell us something about Venter's status for commentators and colleagues alike: he is a figure to be apprehended by means of analogy.

In 2010, when Venter's institute produced the purportedly first synthetic cell, pictured earlier, Christopher Voigt, an MIT synthetic biologist and engineer, provided a whimsical analogy that cut Venter's achievement down to size:

> There's this great computer in the MIT museum. There's this one computer sitting in there and it is the most intricate woven set of wires. It looks like a rug almost but it was hand put together. That represents the last point when one person could sit there with Radio Shack components and build the best computer in the world.[50]

Voigt's assessment captured the broader skepticism of the engineering oriented synthetic-biologists towards Venter's approach to the field. Impressive? Yes. Efficient and scalable? Not so much. Like the sequencing of the human genome, which proved the apotheosis of the genetic sequencing era and in some sense its demise, Venter was, yet again, accused of having brought a paradigm to its highest peak and logical conclusion.

Venter and the engineers may have made for strange bedfellows, therefore, insofar as their aims, approaches, and techno-aesthetics diverged, yet they shared equal billing in a wave of publications that heralded the arrival of synthetic biology between 2006 and 2009. Weeks after publishing his piece announcing the successful completion of the first step in the J. Craig Venter Institute's attempt to produce a synthetic life-form, for example, Nicolas Wade wrote a column introducing the field of synthetic biology more broadly.[51] "Forget genetic engineering," it began. "The new idea is synthetic biology, an effort by engineers to rewire the genetic circuitry of living organisms." Accompanying the article was a cartoon drawing of a formally dressed white family (mother, father, son, and daughter) standing next to a single-family suburban-looking home and sedan, both of which are made out of assorted plants and vegetables with recognizable forms and visually palpable squishiness. Ironically playing up a lack of technological imagination, the cartoon captures well a critical perspective that sees entrenched and hegemonic social norms instantiating their own reproduction through technology. In the piece, Wade moved seamlessly between Venter's genome transplant and the parts-based approach. An assortment of parts-based practitioners contributed their own two cents about the budding field. Among them, almost at the end of the piece, and contributing one quote, was a computer scientists and electrical engineer named Ron Weiss.

The Labs

Ron Weiss ran a lab in the electrical engineering department at Princeton. In his late thirties, he had recently been tenured when I began my fieldwork. Ron's lab's research is neatly and squarely locatable within the engineering approach, with the minor caveat that he is among a growing group of practitioners doing work on mammalian cells, a far more challenging substrate for engineers than the more commonly manipulated bacteria. Though Princeton was not a hotbed of activity for the engineering approach, Ron had come from MIT, which was not only a center but an epicenter of attempts to analogize biological systems to electrical ones, and he had played a key role in one of several origin stories for this particular approach. Ron had been and continues to be involved in writing programmatically about programmability in biology and invested in the production and stabilization of this version of the field. Ron's lab's positioning was therefore somewhat paradoxical and without a doubt occasionally frustrating for its inhabitants, being in some sense part of a center and in another far from it. For the entire duration of my fieldwork in Ron's lab, these frustrations were particularly apparent, and also on the cusp of being resolved, as Ron had received a tenured offer from his alma mater and the synthetic biology mothership, MIT, to which he would eventually move, lab and all.

Michael Hecht's lab was located in Princeton's chemistry department. In his mid-fifties at the time, Michael had had a thirty-year career at the interface of chemistry and molecular biology. Michael arrived at the understanding that his research trajectory had led him to synthetic biology relatively late in his career. His lab's primary research involves rehabilitating sickly *E. coli* using synthetic proteins that his lab engineers using a design technique that Michael had devised himself. The goal of the research is to show that de novo proteins designed for form, not function, can support life. Michael's lab's work doesn't fit neatly within any of the mediagenic paradigms described above, though he questioned their claims to making biological things "from scratch" on the grounds that, whereas all of these approaches shuffled existing genetic sequences, his lab based its approach to synthetic biology on the construction of novel proteins that cannot be found in nature. Given the exemplars above, the question, "Is Michael *really* a synthetic biologist?" is perhaps a legitimate one, though the very asking of it raises questions about the criteria for group membership and belonging in a field in which inclusiveness of a broad range of activities has resulted from a balancing act that pits the pros of shared hype against the cons of a diluted brand.[52]

Incidentally, I conducted my fieldwork while I was a graduate student at Princeton University. Walks to the field were short. In more metaphorical terms, distance varied. Princeton is a token of a type of institution in which the individuals whose research lives are chronicled in this book and I have all spent much of our time: the university. The endeavors this institution links together—synthetic biology among them—share some features, and are at least partly oriented toward knowledge. My experience of researching and writing this book has thus borne out Dominic Boyer's observation that, because it invariably entails "symmetrically turning anthropology inward on its own epistemic practices, forms and relations," the anthropology of knowledge can be "discomfiting: it aches a little to do the anthropology of knowledge just like it aches to do any effective therapy."[53] The ache is certainly there. In parts of the book, I attempt to transmit some of it to the reader by marking out areas of potential cross-disciplinary reflexivity about knowledge production as well as the institutions and practices associated with it. Be forewarned, the effectiveness of the therapy I'm less sure about.

Chapter 1

Labs, Lives, Technoscience

The Engineering Quadrangle (E-Quad) at Princeton University is located at the eastern edge of campus. Constructed in the 1960s, the building's design marks a shift away from the collegiate gothic style that reigns supreme farther to the west. The effect, therefore, of approaching E-Quad's façade from the center of campus is a sense of crossing the threshold to modernity. Up the steps, inside, Ron Weiss could most often be found in his office in the Department of Electrical Engineering, where he had been employed since 2001.

Ron's lab, located on the same floor as his office, occupied two discrete rooms at the end of a long corridor. Both rooms were impeccably neat, owing to the gruff but supportive presence of the lab manager, Mike, whose full-time job was to oversee lab nitty-gritty, restock, and make sure that everything was in working order. Mike was stationed in the larger and older lab, which contained lots of bench space, computers, machines, instruments, desks, flasks, bottles, and refrigerators, as well as a special room for the Zeiss microscope, one of the most expensive pieces of equipment in the lab's possession and the one that required the most resource-sharing coordination between individuals who signed up to use it at all hours of the night. The other lab space, located across the hall, was newer and smaller, and contained all of its own basic equipment, as well as a cold room and a cell culture and virus room with two fume hoods.

Ron directed these laboratory spaces, but the majority of his time was not spent in them. The primary inhabitants of these rooms were undergraduates, graduate students, and postdocs, who brought with them different levels of expertise in a variety of disciplines. Since Princeton had no bioengineering track or program, the more permanent members of Ron's lab were funneled in through either molecular biology or engineering fields. As is often the case in academic laboratories, graduate students early in their careers rotated among a few labs for short test periods, gauging their interest in the research and their compatibility with the setting before settling on the lab in which they would pursue their graduate degrees. A number of graduate students had chosen Ron's lab as their permanent home. The lab also housed four postdocs, representing four different fields: neuroscience, molecular biology, biophysics, and protein chemistry. The biologically trained students mostly lacked familiarity with circuit abstractions and computational models, whereas the engineering students often needed training in wet lab work. Everyone was an amateur at something.

At Princeton, Ron was the only adherent of the parts-based approach to synthetic biology. He was in his late thirties, which placed him on the younger end of a generation of practitioners in the budding field to be running their own labs. Like the older crop of mid-career scientists and engineers redirecting their research efforts and retooling their labs to do synthetic biology, Ron necessarily built an assortment of competencies in markedly unstructured settings. This was one of the defining features of Ron's description of his own career path (which I recount below). This feature resonates more generally with the way the parts-based approach to synthetic biology, pursued mostly by trained engineers, has taken shape among practitioners who not only knew very little about biological systems at the outset but also lacked the more organic know-how tied to the daily care and manipulation of living things in the lab.

In her classic study of physicists, Sharon Traweek tracks the way the experimental particle physics community reproduces itself through the training of novices.[1] She identifies the patterns through which education and inculcation occur, and by which particle physicists learn the criteria for a successful career. The images she conveys of community, stability, and gendered reproduction can be discerned only within a sufficiently entrenched discipline. In contrast, in an unstable and ambiguously bounded field like synthetic biology, idiosyncratic individual paths figure prominently, especially for members of the first generation of practitioners, like Ron and Michael, whose training necessarily took place within—or between—the reproductive mechanisms of established disciplines. Such paths embed different concepts and logics within the synthetic organisms made in different labs. They also culminate in different normative frameworks for assessing what counts as a project worth pursuing, or a question worth asking, or a life worth making.

I interviewed Ron fairly regularly during my time in his lab. On one occasion, when I asked Ron whether he identified as a synthetic biologist, he demurred, noting that the label sounded funny. No, he explained, he identified first and foremost as an engineer.

"I've always been interested in computers," he explained. He had inherited the interest in computers from his father, who had worked for IBM in Israel before moving the family to the United States, when Ron was fourteen years old, to take a job at a software company in Texas.

Ron spent much of his childhood and adolescence programming. When he enrolled as an undergraduate at Brandeis University, he knew he wanted to pursue computer science. In his senior year at Brandeis, he applied to graduate school and, in the early 1990s, joined MIT's prestigious Department of Electrical Engineering and Computer Science as a doctoral student.

At MIT, he began his studies with a focus on digital media and information retrieval, but he had not found research in these areas to be "life-fulfilling" work. Then came amorphous computing: the term was coined in a 1996 white paper, "Amorphous Computing Manifesto," whose lead authors were all faculty in MIT's Department of Electrical Engineering and Computer Science.[2] The term "amorphous computing" describes computational systems made up of very large numbers of processors that interact locally and that possess limited individual computational ability. Researchers studying amorphous computations draw inspiration from biological systems, which provide models and examples of how huge numbers of irregular and computationally limited bits can coordinate behavior or produce global patterns. It was therefore through amorphous computing that Ron first encountered research at the interface of programming and biology. Yet the step from amorphous computing to synthetic biology was still a fairly drastic one. Ron explained it to me as follows:

> I was looking at biology as a way to get inspired for how to program computers, and I was doing all kinds of simulation and things like that. There was a point when I came to the conclusion that rather than wanting to look at biology as a way to get inspired for how to program computers, I actually want to reverse the arrow and say, how can I look at computing and understand how to program biology. And at that point I teamed up with the person who ended up being my main adviser, Tom Knight, who's actually one of the visionaries in synthetic biology. He's one of the people who started the field. And I helped him set up a wet lab in the computer science building. From then on I was working on bioengineering.[3]

Ron was not alone in praising his mentor. In a write-up about synthetic biology from 2005 that appeared in *Wired* magazine, Oliver Morton called Knight "an MIT institution."[4] A computer engineer, Knight spent much of his life in and around MIT's Computer Science and Artificial Intelligence Laboratory. He is often credited with having been in some sense the "father" of synthetic biology, an analogy that draws together both his role in elaborating the basic idea of how biology might become the substrate of choice for a veritable engineering field and his efforts to spur institutional and infrastructural developments that would bring such a field to fruition. Knight, for example, was one of the individuals responsible for launching the International Genetically Engineered Machine Competition (iGEM), an event worthy of a brief descriptive detour, since Ron's lab was peopled by its participants and engrossed in its pursuit for a good portion of the year.

iGEM is a remarkable synthetic biology fête where institutionally affiliated teams of undergraduates tackle synthetic biology projects under the guidance of faculty, postdocs, and graduate students. Celebrated as a major site of infrastructure building for the parts-based approach, iGEM is rife with peculiarly late-twentieth- and early-twenty-first-century technological and institutional arrangements. Corporate sponsorships for individual teams are de rigueur. Logos, public and private, are emblazoned on team T-shirts and posters. The competition was launched in the early 2000s alongside a template for putting together standard biological parts called BioBricks. iGEM teams built, characterized, and circulated these parts, growing a material and informational library for bacterial synthetic biology.[5] Ron had assisted in the early stages and had led a team for Princeton every year since the competition's inception.

iGEM was launched as an attempt to replicate some of the magic of a previous era in MIT computer science, one defined by Lynn Conway's legendary large-scale integration class, which is considered by many to have revolutionized electronics. Conway, a computer architect from Xerox PARC, in collaboration with Caltech's Carver Mead, developed a new chip-making method called VLSI, which separated the design process from manufacturing. Conway's course, first taught in the late 1970s, was a response to the strategic secrecy of early semiconductor companies that used the technology for limited military and industrial applications. It offered a hands-on experience for participants, owing largely to grant support from DARPA that allowed students to have their chip designs assembled at a chip foundry in California. The hands-on ethic, the ideals of openness and free(ish) circulation, and the recruitment of energetic and naïve youngsters as the catalysts for a technological revolution were all features Knight and a few others sought to recreate. Even DARPA did its part at the outset, paying the cost of DNA synthesis.

Before iGEM, while still a graduate student at MIT, Ron helped Knight get started. First, Ron, Knight, and one other graduate student who was aiding in the efforts had to learn how to work with cells. They were completely unfamiliar with wet lab work. They started by taking some undergraduate courses in biology, reading papers and books, and talking to people. Mainly, Ron recalled, "We just picked it up as we went along . . . and just tried things. It was probably not the most efficient way to learn, but it's kind of the MIT way." The institution, Ron claimed, was supportive of innovative work and propagated an ethos of self-reliance (a characterization to which I return in chapter 2). "At MIT they are very open. They like crazy ideas. Doing something nontraditional is something that people enjoy. You pretty much assume that you can figure it out by yourself and you don't need anybody else."

Having gotten the wet lab going, Ron set about running experiments. As he recalled, the contrast between spending one's days programming and doing wet lab work was a stark one. Whereas in computer science, "if you make a mistake you can go back and fix it, in biology, it doesn't work that way." Mistakes now meant having to repeat specific protocols or entire experiments. One had to be diligent and attentive, Ron explained, remembering the frustration of those months, and the doubts that accompanied the daily lab work: "I was thinking to myself, maybe I shouldn't have switched to do all this biology stuff."

Ron soon encountered the first real hurdle when he tried to build a plasmid. Plasmids are circular pieces of DNA that replicate independent of chromosomes and that allow experimenters to manipulate genes. Plasmid building is notoriously tedious and prone to human error, increasing the temporal and energetic start-up costs of research drastically. On the surface, plasmid building requires that practitioners follow a fairly straightforward step-by-step process. Yet discussions of the fickle outcomes of this process were often fringed with an air of mystery. Some practitioners had consistently better luck for unknown reasons. Indeed, as Ron recalled, it took him six whole months to confirm his first plasmid. A week later, a molecular biologist the lab had hired for assistance arrived on the scene. In recollecting her effect on their work, Ron implicitly recanted some of his enthusiasm for sacrificing efficiency in the name of self-reliance. He realized the value of this new lab member quickly: "She was a biologist. An actual biologist. Which is something we should have done week one. We should have had an *actual* biologist."

That first plasmid was put to use in a digital logic circuit made of cells, which served as a major part of Ron's doctoral thesis. The first generation of circuits, like the ones Ron built, was used to exemplify the utility of engineering abstractions for the construction of novel biological systems. Thus, these early, rudimentary circuits achieved fairly modest technological goals that worked to validate the approach while also cinching a set of engineering abstractions to a new substrate. Referring to this first generation of circuits, a collaborator of Ron's in molecular biology at Princeton once remarked that it was neat that engineers had taught a dog how to talk. He quickly added that it was time to start caring about what the dog was actually saying.

Ron spent nine years at MIT. He didn't want to leave, but his family had grown, and "it was time for a real job." He went on the academic job market, applying, mostly, to computer science departments. Princeton's was in fact the lone electrical engineering department in the mix. His interview went well, despite some perceived resistance from the molecular biologists who came to vet his job talk and who, Ron sensed, unlike engineers, didn't quite see the point of the work. Nonetheless, Ron was offered the job at Princeton and accepted it. A subsequent steady stream of well-placed publications secured his tenure case and his standing among practitioners of the parts-based approach.

Michael Hecht and the members of his lab didn't talk about circuits or wiring. None of them had backgrounds in engineering. Yet Michael's research emerged from a tradition of work that provided epistemic validation for contemporary attempts to found a biological discipline aimed at synthesis.

In 1828, Friedrich Wöhler famously synthesized urea, an organic compound secreted in urine, shattering the divide between the organic and the inorganic. At the time, scientists had managed to cross the divide in only one direction, transforming organic molecules into inorganic ones through various treatments. But the inability to perform the opposite operation had bolstered the notion that some vital force divided the animate from the inanimate. In the decades that followed Wöhler's discovery, chemists began not only to assemble an array of organic molecules but also to synthesize new compounds similar to those found in nature, contributing to modern theories of chemical structure and reactivity, while also leaving their mark on many aspects of human life, from medicine to agriculture and beyond.[6] The notion that synthesis and analysis could be conjoined was therefore not a novel insight for a chemist, nor was the deployment of this particular pairing at the boundaries of the living.

In his mid-fifties, with booming voice and jovial charisma, Michael had spent much of his career as a professor in Princeton's chemistry department. When I began my fieldwork, his lab and office were located in Hoyt, a 1970s extension to the gothic Frick Chemistry Laboratory at Princeton, which was conveniently located one door down from the anthropology department on Washington Road. (Since I recount the story of the lab's move to the new home of the chemistry department in chapter 5, I leave the space descriptively barren for now.)

Michael's lab was inhabited by protein chemists: four graduate students from the chemistry department and two postdocs. Another chemistry graduate student joined the lab midway through the year, a casualty from one of the largest and by all accounts most cutthroat labs in the department. Whereas Ron's lab's inhabitants were pursuing assorted projects, Michael's lab mostly directed collective efforts toward two projects: one involving testing synthetic de novo proteins for their ability to support life and another involving Alzheimer's disease.

When I asked Michael how he had come to view his work as a contribution to synthetic biology, he quipped, "I wanted to be a shrink." The son of German Jewish immigrants, Michael grew up in New York City. He attended New York City schools and then went on to Cornell for his undergraduate degree. There, he took part in the college scholar program, which freed him from having to choose a major or fulfill university requirements. Having discovered that the psychology department didn't offer a clinical path, he decided to take a break after his sophomore year. He hitchhiked around the country for a while, sending his Cornell undergraduate adviser the occasional postcard. In November, the cold swept in and he abandoned his tour, returning to Cornell. The fall semester was already under way, and the spring semester a long way off. "It was flipping burgers or lab work." Michael's undergraduate adviser called everyone he knew in the chemistry department. Finally, the adviser contacted Harold Scheraga, considered a pioneer in the area of protein biophysics. Scheraga agreed to give Michael a job in his lab, where Michael helped with research at the interface of chemistry and biology. Michael recalled enjoying the research, though he still wasn't sure what he wanted to do for a living, until he read James Watson's textbook, *The Molecular Biology of the Gene* and became fascinated. He described the book as a precipitant in his decision to go to graduate school in molecular biology, though he quickly added, "I really was pretty darn clueless."

Like Ron, Michael went to MIT and, eventually, joined the lab of a young professor who was doing early work in protein folding and protein engineering. "I had a really good graduate experience," recounted Michael, who, also like Ron, had overwhelmingly positive memories of MIT, though Michael attributed his appreciation for those years to the particular circumstances in which he found himself, rather than any general sensed qualities of the institution. As he put it, his adviser "was amazing, and the project was amazing."

After graduate school, Michael applied for a postdoc with David and Jane Richardson at Duke. One of Jane Richardson's claims to fame, beside the fact that she had never received graduate training in the sciences but had succeeded in them spectacularly nonetheless, was that she originated the now ubiquitous Richardson diagrams, or ribbon diagrams, for proteins. Michael was awarded the postdoc, and so, after traveling around the world, he landed in the Richardson's lab, where he helped them build de novo proteins from scratch by mutating amino acids.

Michael used the same word to describe Jane Richardson that Ron used to describe Tom Knight: "visionary." In these cases, the term, which endows the sage prophet or the wise master with the virtue of imagination and foresight, produces a particular picture of research and discovery, driven by those who have the rare ability to see ahead, though Michael's use of the term "visionary" was also literal, as he quickly clarified: "Jane . . . could actually visualize [proteins] as these ribbon diagrams. She could visualize the whole structure. And people used to say that the protein database was a computerized version of Jane."[7]

After completing his postdoc in the Richardson's lab, Michael was recruited to the chemistry department at Princeton. His most prominent early career work, which culminated in a paper in *Science*, was the invention of a technique for building libraries of stably folding proteins.[8] The technique involves arranging amino acids in specific configurations of polar and nonpolar elements, which allows the proteins to fold and hold their structure. The building blocks of the actual proteins within the abstract pattern of polar and nonpolar elements can be varied, creating a vast number of proteins that follow the same patterns of polarity but have different amino acid sequences plugged in.

Michael's technique for producing libraries of stably folding proteins was included in a textbook the year after the technique was published. It was a significant achievement, on the heels of which he received an offer from another university. He went up for early tenure at Princeton, with some assurances of a positive outcome from department members, and was denied. This was the first in a series of disappointments stemming from an altogether miserable tenure process, which dragged on for years and reflected a department that Michael characterized as having been, at the time, "a mess." Before the tenure debacle, Michael explained, "it was all just falling into things, it really was: undergrad, grad, postdoc, faculty, until all of a sudden, when tenure came around, it didn't fall into place." Eventually, it did. The years since had allowed him to focus more fully on research, though the protracted tenure process had left its mark on his relationship with his home department (a subject I return to in chapter 5).

Whereas Ron's understanding of his own arrival at the idea of synthetic biology involved a "eureka" moment, a conceptual reversal, and a sudden change in the material infrastructure of his research, Michael's research, in his own estimation, had traced a logical progression of questions that build naturally on one another. At Cornell, he had done protein folding in the lab of Harold Scheraga, and then at MIT he had experimented with protein mutagenesis. "With Bob [at MIT]," explained Michael, "we had mutated lots of amino acids. What's the next conceptual step? You mutate every amino acid, i.e., design from scratch. So that was a logical step. So I did that with Dave and Jane Richardson." With the Richardsons, Michael had designed one protein at a time. Next, Michael's homegrown technique allowed him to design a whole library of proteins that folded predictably. The next question, for Michael, which he framed as following naturally, was whether these binary coded proteins could function in an organism. He explained:

> So you do all sorts of biochemistry experiments to see if they can function. We did that. And then, what I always knew because I trained in Bob's lab at MIT as a biologist, I said, well, all that doesn't matter, what really matters is, Do they function in vivo? And so for a long while we were trying to get them to function in vivo by taking a particular strain [of bacterium] that was defective for a particular activity and getting it to rescue. And then one day I heard a seminar somewhere where somebody mentioned this Keio collection, where you can get all the strains that are deficient for all the activities, and I said, that's the way to do this, because if we're trying to rescue one thing, it may or may not work, but if you have four thousand to try, something's going to work. And so that's what we did.

The Keio collection comprises a set of *E. coli* auxotrophs: mutated *E. coli* strains that can grow when given rich media (the nutritional equivalent of chicken soup) but cannot grow on poor media (essentially sugar water). Michael's lab then tested their own proteins for their ability to compensate for the deleted functions in the auxotrophs. They managed to "rescue" four *E. coli* strains, thereby demonstrating that their own novel proteins, which bear little resemblance to natural sequences, can enable life.

Though Michael's lab's proteins managed to save the sickly bacteria, at the time of my fieldwork, lab members did not yet know how this feat had been achieved. Without the account of how their proteins worked, lab members were having trouble satisfying the disciplinary demands of chemistry, in which interest in function outpaced appetite for brute technological achievements (I return to this problem in chapter 4). Lab members did know that their proteins, which all take the form of four helix bundles, were small and structurally simple compared with the naturally occurring proteins that normally do the work of supporting life. Keeping this observation in mind, and with some knowledge of the research pursued in Michael's lab, we can now submit another interpretation for Michael's claim that "making life is easy" (discussed in the introduction). The phrase can be understood through the lens of Michael's lab's research. A provocation of sorts, the claim is aimed at the miracle of creation, synthetic, divine, or Darwinian. And the logic is as follows: If comparatively small proteins designed for form rather than function can support life, then you could throw a potentially wide range of ingredients into primordial soup. You don't even have to design the ingredients rationally. A smidgen of design to make sure they fold, some screening assays, and—poof!—life.[9]

When synthetic biology came along in the early 2000s, Michael recalls, "I said, oh, we're doing that." He quickly adds, "But we're doing more synthetic biology than they are"—a phrase that Michael repeated on occasion and that simultaneously marks out his lab's research as different from the more centrally located efforts and asserts its importance in that milieu. In part, "we're doing more synthetic biology than they are" critically chastises adherents of the parts-based approach for using genetic sequences given in nature and then claiming such repurposing as the basis of biological engineering "from scratch." If you want parts from which to build life "from scratch," Michael explains, you have to create novel genes that code for novel proteins. This, for Michael, was the true horizon of the biology of the possible.

Venter's strong presence in the synthetic biology hype sphere also notably affected morale in Michael's lab, more so than in Ron's. Whereas the parts-based approach easily claimed critical distance from Venter's synthetic genomes, Michael's lab's research was less able to shield itself through an existing alternate collective paradigm. Michael therefore articulated important similarities and differences between his lab's work and Venter's in his criticisms of Venter's approach, while also making clear that the relative positioning irked him. In his view, Venter and his team had failed to build something "from scratch" as well. They had merely replicated what already existed. All of the cells produced by the Venter transplants mimicked existing life-forms. Michael therefore found Venter's signature achievement to be much less significant than its media coverage would suggest and questioned the merit of

Venter's claim to have made "synthetic life" in relation to Michael's own, which was grounded in de novo proteins with no natural analogues. In a talk in late 2010, Michael argued the point:

> Several months ago, in the spring, there was this big press release and big to-do when Craig Venter and colleagues came out with what they called the creation of an artificial cell. That was copying natural sequences and placing them into a cell. It was synthetic in the sense that they knew the information from biology, they copied it, and they put it in the cell. Not really synthetic biology, in my view.[10]

Their own project, Michael said, "is in principle way more profound than what Craig Venter is doing because we're using things made from scratch." But he was also jaded by the difficulty of convincing others of the significance of the work. "I'm not sure they'll see that," he appended. When Michael found out the exorbitant price tag of Venter's research, he announced to his grad students and postdocs, wryly, "You are all underpaid." Reflecting on a key paper detailing his lab's findings (the publishing saga of which I analyze in chapter 4), Michael lamented, "It has taken forever to get that paper out, which has been kind of painful. Obviously, it's hard to be distant from one's own work. It's been frustrating."

Michael had been drawn into the synthetic biology "center" on occasion. He had, for example, participated in something called a "sand pit," a venue put together by a prestigious grant organization for a handful of synthetic biologists to come up with new ideas in teams and then pitch the ideas for funding. As he watched interest follow power and prestige, Michael's enthusiasm waned. Turns out, he explained, sand pits will often beget the same behavior: eventually the kids will throw sand in each other's faces. The taunt "we're doing more synthetic biology than they are," then, also registers something of a grievance. It contests the grounds on which some things get to matter, and belong, more than others.

The Normalization of Technoscience

Synthetic biology is often described as a model technoscience. In this regard, the paths discussed above are also stories about the pursuit of technoscience as a contemporary academic vocation that mixes knowledge and intervention as means and ends. The same can be said of many of the episodes and stories I recount in later chapters. Yet such a framing requires some specification. "Synthetic biology," as a term, stands for too many sorts of activities to neatly anchor a unitary understanding of technoscience. At the same time, "technoscience" has served as an equally fuzzy and heterogeneous label that can muddy as much as it clarifies. Thus, relating the two requires some conceptual bootstrapping.

Technoscience is a commonly invoked term in the social sciences and humanities, but it indexes a number of different dynamics, debates, and questions. At a certain remove, technoscience refers to different kinds of mixtures and mash-ups that defy the ideal typic division of labor between activities aimed at representation and those aimed at technological intervention. Such mash-ups can be organized around two broad and conflicting positions. The first implies that the entwining of representation and technical intervention is a hallmark of *all* modern science. The second suggests that technoscience is a more recent development, separable from a properly scientific age that predates it.

As Evelyn Fox Keller recounts, the term technoscience was coined in the 1970s, by the philosopher Gilbert Hottois, and swiftly taken up by a wave of prominent science studies scholars—herself, Bruno Latour, and Donna Haraway among them—as a means of questioning the distinction between representing and intervening, the pure and the applied.[11] More specifically, Latour, Haraway, and others were targeting the pure side of the equation by asserting that facts are always, in an important sense, made, not discovered, and therefore inseparable from politics. The notion that the modern sciences are predicated on intervention, that facts are made not found, lies behind the term's initial popularity. Technoscience "was a manifestly polemical intervention," recounts Keller, "aimed at debunking the myth of pure science, and recommending that henceforth we will not speak of science and engineering (or of science and technology), rather, we will speak only of the real world of technoscience."[12]

If these original deployments of technoscience operated polemically, argues Keller, the polemical edge has mostly faded. Partly, we may note, the normalization of the label might reflect the way constructivism itself has shifted from its original role as a deflationary critical discourse to a generally accepted assessment of the way things really are for those raised in the trenches of the science wars.[13] The notion that facts are made has thus gained the status of a truism in certain corners of the humanities and humanistic social sciences.[14] But this is not Keller's primary claim. Her argument, instead, is that technoscience has lost its polemical edge because of changing contemporary institutional dynamics: "Our institutions are reorganizing into new configurations, traditional disciplinary boundaries are blurring, and the classical divide between science and technology is less and less meaningful to new generations of practicing scientists."[15] Keller's, then, is primarily a sociological and historical claim. Its main point is that, if we look at the salient settings, we find an increasing disregard for the proper separation of science and engineering. Keller's example of the institutional reorganization that has blurred the divide between the pure and the applied is the tenure case of Frances Arnold, in the Division of Chemistry and Chemical Engineering at Caltech. Arnold's lab uses directed evolution to generate enzymes and organisms for various applications. As Keller recounts, "[Arnold's] tenure dispute revolved around the question of whether this was science or not, a question raised by (at least some of) the senior members of the committee. The junior members, who apparently prevailed, failed to see any basis for such a question."[16] The larger scientific community seems to have aligned with these junior members. Among the many forms of professional recognition Arnold has received over the course of an illustrious career was the 2018 Nobel Prize in Chemistry. At issue, then, are shifting norms that dictate the proper balance between knowledge and technological intervention in different academic institutional settings, and a related set of values that defines legitimate means and ends.

Keller is among many scholars who argue that technoscience, today, describes a novel configuration of fairly recent origin.[17] The philosopher of technoscience Alfred Nordmann similarly makes a case for an epochal break between what he terms "the scientific enterprise" and "the regime of technoscience."[18] The term "scientific enterprise," for Nordmann, is on par with "Enlightenment," or "modernity." Each describes a common pursuit indissoluble from a worldview. Nordmann suggests that, as part of the "scientific enterprise," the sciences have historically been committed to a quest for truth, where truth might be seen as a *telos*, or "regulative ideal."[19] The overarching "scientific enterprise" has consequences for epistemic values. For example, scientific notions of objectivity, observes Nordmann, entail at least an effort to separate out the enduring from the contingent, to distill eternal truths from changeable contexts.

To characterize the "regime of technoscience," Nordmann borrows from Bruno Latour's conceptual tool kit. Latour famously argued that the West has never been modern because it has never quite been able to separate nature and culture.[20] Yet undertaking the work of "purifying" nonhuman nature from human culture, despite its futility, defines the modern age. It is precisely this work of purification that the regime of technoscience abandons altogether in Nordmann's view. The entanglements between nature and culture, science and technology, have simply become too obvious to ignore. An "in silico" experiment performed on a computer model and experimentation with genetically engineered animals suggest the same thing: "In both cases it becomes impossible, as a matter of fact, to conceptually determine where human intervention ends and the purely natural process begins."[21] This is because such human artifacts "exhibit properties and processes that are themselves engineered—their relevant dispositions are aspects simultaneously of nature and culture."[22] If purification is no longer possible, it is also no longer *required* in the regime of technoscience. Nordmann explains, "Because these experiments serve mostly to demonstrate practically achieved control of the phenomena, there is also no need to determine this—the achievement stands on its own and validates itself."[23] Such a shift can be observed in different corners of experimental research, where building new technological artifacts and demonstrating practical control have become increasingly acceptable research goals.

Thus, one version of technoscience takes as its object an institutional and normative reorganization, grounded in ever-increasing entanglements between the natural and the contrived. The rapid spread of synthetic biology to different academic settings has certainly benefited from such institutional and normative shifts. Yet it is worth noting that the aim of demonstrating practical control still entails making and judging truth claims. It also requires that technoscientists continue to disentangle results from the contingencies of context, not in order to decipher eternal truths, but rather in order to convincingly claim the requisite agency over their experiments. We may therefore observe much continuity between something like the "scientific enterprise" and the "regime of technoscience" at the level of practice.

To situate the engineering approach to synthetic biology in relation to technoscience, we proceed not from the pure to the applied, but from the applied to the pure, taking into account the demise of what has been called "the linear model," the idea of a kind of knowledge supply chain that goes from the pure to the applied, and on which much discourse in favor of basic research has been based. [24] The linear model has been heavily criticized since the 1990s, its one-way flow replaced by attention to intricate and heterogeneous links between knowledge and technological innovation.[25] The demise of the linear model proves useful for thinking about twentieth- and twenty-first-century biological sciences and their relationship to the engineering approach in synthetic biology. In the mid-twentieth century, the unity of all living things was thought to lie in molecular processes. The models that purportedly described these processes placed genes at the center of the action and suggested elegance and simplicity, placing the possibility of "decoding" life within reach. Hence, the Big Science endeavors of the 1980s and 1990s, with the Human Genome Project their exalted pinnacle. And with these endeavors came data, data, and more data—data that in some sense undid the hoped-for simplicity. Biology is today in the midst of both a "complexity explosion," and a "theory crisis," its focus—life—more enigmatic than it had seemed in previous decades. In come the engineers, who have argued that the combination of synthesis and engineering knowledge practices might do a lot to organize biological data and generate theory. Thus, engineers turn out to be technoscientists too.

If such characterizations of technoscience aim at generality, the ethnographic perspective grounds some different insights into the relationship between knowledge and intervention in emergent experimental settings. Throughout this book, I highlight a more fine-grained dimension of experimental life in contemporary technoscience, a dimension that becomes visible once we view research as an ongoing process. A focus on this temporal dimension reveals that the relationship between knowledge and intervention is not always fixed, but rather unfolds in the ongoing, tentative articulation of links among problems, experimental systems, and new biological things. In chapter 2, I introduce the notion of an "epistemic wager" in order to articulate the element of conjecture at the heart of synthetic biologists' attempts to refigure relations between making and (not) knowing, and I argue that practitioners adjust relations between knowledge and intervention, as both means and ends, on the go.

We visit one last scholar on technoscience who also puts the term in its place. In his fascinating study of the cultural significance of science and technology, Paul Forman argues that science, once thought to be the source of technological change and the pinnacle of cultural value, has ceded ground to engineering.[26] For Forman, this reevaluation signals an epochal shift, though not the one often referenced in writings on changing relations between science and technology, which often hold that something happened to the primacy of science on the way to The Bomb. Never again, this story goes, could science be divorced from technology, or from politics, for that matter. Foreman dismisses this version of events, insisting that science emerged from World War II stronger than ever. It was the rejection of procedure as the guarantor of truth and the ascent of a pragmatic utilitarianism that ranks ends over means that marked the shift, in Forman's account, and characterized a "postmodern ethos." Forman explains,

> In modernity, the cultural rank of science was elevated by that epoch's most basic cultural presuppositions—not merely the presupposition of the superiority of theory to practice, but more importantly the elevation of the public over the private and the disinterested over the interested, and, more importantly still, the belief that the means sanctify the ends, that adherence to proper means is the best guarantee of a "truly good" outcome. Today, on the contrary, technology is the beneficiary, and science the maleficiary, of our pragmatic-utilitarian subordination of means to ends, and of the concomitants of that predominant cultural presupposition, notably, disbelief in disinterestedness and condescension toward conceptual structures.[27]

Forman does more than claim that the hierarchy of scientists and engineers has been reordered. He argues that this reranking signals a broader reorientation in the cultural milieu, which he sums up this way: "Modernity is when 'science' denotes technology too; postmodernity, when 'technology' denotes science too."[28] This is because "technology has now replaced science not only as the principal model for knowledge production, but has also replaced science as principal model for all those 'ordering' activities that constitute culture."[29]

Forman isolates two sites as places to look for this shift. The first is the small academic lab, which he contrasts with the modernism of the "big science" particle physics accelerator labs. Forman explains: "If we want to see postmodernization working radically and admittedly upon the scientific role and knowledge production, we should look . . . into the 'little science' academic laboratories that have reoriented themselves most completely toward technologically defined ends, while it remains to the earlier much-maligned 'big science' laboratories to shoulder the burden of sustaining 'for its own sake,' 'fundamental,' 'curiosity driven,' or 'pure' science—all equally depreciated epithets."[30]

The second site Forman isolates as a diagnostic is more surprising. It is the abandonment of the divide between science and technology in science studies. The underspecified hybrid amalgam "technoscience," for Forman, expressed a still fundamentally modern rationality when it gained popularity in the 1980s, showing that the "conflation of technology and science initially proceeded not from any especially high valuation of technology, but rather from a still modern preoccupation with science, and a still more old-fashioned disapprobation accompanying the newly arisen recognition of science's manifold entanglements with technology."[31] The reconceptualization of science studies as a field concerned with the hybrid technology/science that followed suit took place in the 1980s, and has continued to reign supreme in science and technology studies, according to Forman. The shift, argues Forman, was too total, quick, and silent to have been a reasoned response to a changing empirical situation. Hence, he locates the impetus for it in the cultural ether. Perhaps, then, the normalization of technoscience as a description of the way things really are rather than polemic constitutes, in Forman's epochal terms, a postmodern usage of the term that has come to replace a modern one. What STS presents us with, in that case, are a set of postmodern studies of postmodernity. Forman's highly original work is thus a reminder of the way analytics are also products of their times, their sudden popularity often understandable not only in terms of the territory they map or the problems they solve but also as a sign of the cultural milieu of which they form a part.

Chapter 2

The Virtues of the Naïve View

As I noted in the introduction, life is easy to make today. Yet difficulties abound. For one, biology, many have argued, is significantly more complex than was anticipated before the advent of modern sequencing technologies. In this chapter, I take the destabilization of knowledge in the postgenomic era to mean that the question concerning how *not knowing* will shape the so-called century of biology is as pressing as the one that takes relations between knowledge and intervention as its starting point.

We zoom in on a particular scene in Ron's recollection of his own involvement in launching synthetic biology at MIT, from chapter 1. Recall that, in his account, Ron included the observation that he, Tom Knight, and the other graduate student who was engaged in setting up the wet lab at MIT "just picked it up," as they went along. "It was probably not the most efficient way to learn, but it's kind of the MIT way. You can figure it out by yourself and you don't need anybody else."

The statement seems a bit mysterious when we interrogate it more closely. The engineers' criticisms of Venter's genome transplant, mentioned in the introduction, were often couched in terms of Venter's method's inefficiency. In these critical appraisals of Venter's work, the value of efficiency, from an engineering standpoint, was taken to be self-evident. Why, then, would Ron and his engineering peers forgo efficiency for self-reliance at the founding moment of the parts-based approach to synthetic biology? Attitudinally, why would this inefficiency be a matter for even a hint of celebration?

One possible answer is that Ron's endorsement of MIT's do-it-yourself ethos squares well with an engineer's pragmatic privileging of practical knowledge over theoretical knowledge. That many trained engineers pursuing synthetic biology at least partly subscribe to this ranking is evidenced by a slogan that many of them have recruited to explain their own approach to building biological systems: "learning by doing." The slogan is routinely invoked among parts-based practitioners. Yet, if we read the slogan not for what it affirms, but for what it negates, does an endorsement of "learning by doing" subtly imply a refusal of book learning? Does it suggest the cultivation of a certain kind of ignorant stance at the outset?

To give this stance some empirical form, I submit a piece of evidence. James Collins, a trained medical engineer and prominent synthetic biologist, originator of the groundbreaking genetic switch described in the introduction, mythologizes "naïveté" among an early crop of parts-based practitioners. In several lectures, Collins repeats a well-rehearsed anecdote about the events leading up to the implementation of the genetic switch. The story varies a little from performance to performance, but the basic outline remains fixed. Collins recalls visiting Charles Cantor, a highly regarded molecular geneticist, eventual second author on the "switch" paper and then chair of bioengineering at Boston University. He says,

> Tim Gardner and I went to a bunch of different molecular biologists we knew, to see what do they think. And so, this is 1998, we print out our transparencies, we go over to their offices, we share this, and almost every conversation went as follows. . . . Aw, very cool. Ooh. Very interesting. Do you think this can work? No. And we press 'em. Why not? Well, you know, leakage. . . . You're looking at very large plasmids. The system's not gonna like that. They're not gonna like having this in them, they're gonna mutate, they're going to spit it out, et cetera. And invariably they would end the conversation by patting us on the head and saying, you know guys, biology is complicated. Why don't you stick to engineering? And, I'll come back to that, they were right in many ways, but we were kinda discouraged. We went to Charles Can-

tor, and said, Charles, what do you think? He says, I got two answers for you guys: one, crazy enough idea I think it can work and, two, you guys are so naïve, I bet you guys are the two that can get it to work. And with that endorsement, which I asked him not to put in my tenure letter at that time, he actually gave us a little bit of space in his lab.[1]

Collins delivers this part of his talk at lightning speed as an engaging aside, heroic yet trivial. The story highlights the sensed skepticism of molecular biologists, phrased in terms of a territorial dispute. Biology here. Engineering there. The biologists' skepticism is founded in familiarity with organisms, evidenced not just by knowledge, but also by anthropomorphic proximity: "they're not gonna like having this in them." Meanwhile, Collins's naïveté, his lack of familiarity with molecular biology, was seen to confer some kind of advantage, fueling audacity in well-trodden soil.

Borrowed from French, the noun naïveté is derived from the Latin *nativus*, which means innate, from birth, natural. A dictionary definition paints three portraits of the naïve: (1) "showing a lack of experience, wisdom or judgment"; (2) "natural and unaffected"; (3) "of or denoting art produced in a straightforward style that deliberately rejects sophisticated artistic techniques and has a bold directness resembling a child's work, typically in bright colors with little or no perspective."[2] The first portrait is the one that gives Collins's anecdote its humorous qualities as a show of self-deprecation. The second is evocative as well: in directing us toward the "natural," the "unaffected," it suggests that engineers approach biological systems without the preconceptions of those who are the heirs and keepers of bodies of knowledge and traditions of thought pertaining to living things. The irony of the "natural" perspective being the precondition for the making of synthetic life seems worth noting. The third leaves its mark as well, in productive tension with the second. The comparison with the domain of artistic production is helpful. Naïve art, too, is stylized. The "as if" innocent perception of the child (or the engineer) is often but a ruse for a rejection of one style in favor of another (a theme I return to in chapter 3).[3]

Drawing on these interpretations, here, I argue that practitioners of the parts-based approach to synthetic biology have tested something we might call "methodological ignorance."

This chapter's intervention thus joins a growing body of scholarship aimed at troubling the view that ignorance is merely the flip side of knowledge, or a shapeless void.[4] Linsey McGoey has argued that studies of ignorance stand to effect "a subtle shift in the epistemological gaze that seeks to offer non-knowledge its full due as a social fact, not as a precursor or an impediment to more knowledge, but as a productive force in itself, as the twin and not the opposite of knowledge."[5] I contribute two main insights to the project of capturing ignorance in its positivity. The first is that ignorance cuts networks and creates epistemic barriers. In framing ignorance as a producer of epistemic breaks and barriers, I invite the reader to think analogically and more generally about the additive models of knowledge that underlie many of our collective fantasies about interdisciplinarity. Having accepted a perspectival view of knowledge, we sometimes hold out hope that gluing perspectives together will give us a handle on phenomena that seem to escape our disciplinary grasp. Here I argue that frictions and barriers might be just as important as additivity, especially when the line between epistemic practices and their effects becomes blurred.[6] The second insight is that, while complexity often seems to invite more knowledge as a response, ignorance can furnish a response to complexity. As Gaymon Bennett has noted, "References to complexity have become a hallmark of critical evaluations of synthetic biology."[7] These critical evaluations stake their skepticism on the idea that the biological substrate is too complex for the rational design approaches of engineers. During my fieldwork, I was struck, therefore, by the way many of the synthetic biologists I talked to or read about, especially those pursuing the parts-based approach, responded to such critical appraisals by knowing less, not more. Drawing on a cross-disciplinary literature, I tease out some of the logics behind this epistemological wager. The idea of an epistemological wager here is important. It acknowledges that knowledge practices can contain an element of risk and conjecture, their outcomes a matter of significant uncertainty.

The reading of synthetic biology's techno-epistemic practices I produce in this chapter is a sort of reverse "hermeneutics of suspicion."[8] Literary theorist Rita Felski describes the hermeneutics of a suspicion as "a distinctively modern style of interpretation that circumvents obvious or self-evident meanings in order to draw out less visible and less flattering truths."[9] In this chapter, I circumvent the obvious and self-evident, but the interpretation I produce is in some senses exaggeratedly flattering when compared with some simpler alternatives. In my approach to ignorance, I make method of what is usually viewed as a deficiency.

The story I tell in this chapter is also a temporal one. Whether epistemic wagers ultimately produce results is an experimental question, one that often gets answered over time. In this case, as time passes and biological systems continue to humor only moderately the interventions of engineers, biological knowledge—and complexity—come back into the picture. This is perhaps why, in 2014, some years after his work on the original genetic switch, Collins provided commentary far less celebratory of the salutary effects of the engineer's naïveté. When *Nature* asked a group of elite synthetic biologists to weigh in on what the field needs in order to progress, Collins responded, "Bring in the biologists."[10] In so doing, Collins conceded that, despite the rhetoric about the essential interdisciplinarity of the field, synthetic biology had been pursued in relative disciplinary isolation. As Collins explained, "Synthetic biology is often described as bringing together engineers and biologists to build genetic circuits for some useful task. In fact, the field has engaged relatively few biologists. This is holding back its progress. We do not yet know enough biology to make synthetic biology a predictable engineering discipline."[11] Collins continued, "Synthetic biology projects are frequently thwarted when engineering runs up against the complexity of biology."[12] As I will argue, this pairing of complexity with the need for a more interdisciplinary approach marks a shift in strategy, one spurred by experimental outcomes.

How Much Biology Do You Need to Know?

Owing to a quick and trivial-seeming encounter with a graduate student, my fieldwork in Ron's lab more or less began with the question, "How much biology do you need to know to do synthetic biology?" Princeton's iGEM team had just taken up residence in Ron's lab. I had attended my first lab meeting and had been introduced to the group. A few days later, on a particularly sticky summer day, I ran into Josh, one of the engineering students pursuing a PhD in Ron's lab, in the parking lot of the university housing complex in which we both lived. Josh was an applied physics major from Caltech who was taking his second-year qualifying exams in electrical engineering. He had thus completed the bulk of his graduate training in the engineering school, unlike some of the other students in Ron's lab, who found their way to synthetic biology through molecular biology and were therefore meeting the milestones of a different disciplinary track. Exchanging some pleasantries as we crossed paths, I mentioned to Josh that I was brushing up on molecular biology in order to follow the day-to-day goings-on in the lab. "You don't have to know much biology to do this stuff," he responded.

To make some sense of Josh's proclamation, it helps to keep in mind that the wedding of biology to a rational design approach has yet to prove wholly successful. In the lab, this meant that projects worked in fits and starts. Lack of knowledge of the systems at hand was often considered a potential culprit in experimental failures. As I describe later in this chapter, lab members learned a lot of molecular biology, particularly at impasses. It was therefore certainly not empirically self-evident that "you don't have to know much biology to do this stuff." Josh's proclamation indexes an ethos more than it furnishes good practical advice.

This ethos, it should be noted, was by no means uniform across the lab. Meg, for example, the lab's key postdoc whose background was in neuroscience, often insisted that intricate knowledge of the biological substrate should be integrated into the projects of synthetic biologists, rather than ignored, a stance that Ron endorsed. At the time of my fieldwork, Meg and Ron were writing a review article together, which led them to discuss the role of biological knowledge in synthetic biology frequently. Since I had offered to assist in editing the paper, I listened as they hashed these issues out. They concurred that, ideally, knowledge of the systems in question should be integrated on the ground floor, at the design stage.

The broader contrast between Meg and Josh is worth dwelling on for a moment, a contrast that is in part dispositional, perhaps disciplinary, and most certainly gendered. Whereas Josh was the youngest member of the lab, excepting the undergraduates, Meg was the oldest (older than Ron, in fact) and most experienced among the members of Ron's lab in the cultivation of cells, an activity she engaged in through idioms of care. She referred to the products of her careful cultivation as "Meg's cells," and "my babies," extending to them a maternal logic that was also a big part of her identity in the lab. She was a parent to three children about whom she spoke often and lovingly, the oldest of whom was now a junior in high school and the youngest member of the iGEM team. Meg also had a chicken coop in her backyard in which she cared for often disabled hens rescued from research. If Josh was self-styled as an engineering student who happened to work with biological systems, Meg was self-styled as a biologist-mother. The instrumentality of the engineering approach therefore sometimes seemed at odds with her participation in particularly maternal discourses of care and cultivation. Her insistence that biological knowledge be integrated at the ground floor could be read along these lines as well, where knowing is a way of being attentive, and ignorance becomes signified as neglect. Perhaps unsurprisingly, Josh and Meg didn't see eye to eye on much of anything.

Yet even Meg framed a key experimental success in her research in terms of her ignorance as a trained neuroscientist working with a different biological substrate. In her account of her research in Ron's lab, her lack of familiarity with both cellular differentiation processes and endocrinology had played a central—and by no means negative—role. Moreover, Meg's project had begun as an iGEM team project, and had thus drawn heavily on the contributions of inexperienced undergraduates. In this sense, the competition, too, offers an interesting site in which to think about the uses of ignorance.[13] Fittingly, when I attended the iGEM jamboree, a team leader at another university extolled the virtues of "virgin minds." These "virgin minds," far from contributing to one-off science fair projects, were building the supposed infrastructure for the engineering approach to the budding field. This was true on the grand institutional scale with the circulation of genetic parts, but it was also true on a smaller one: in Ron's lab, iGEM team members launched research projects, like Meg's, that would far outlive—metaphorically and literally—the students' time in the lab.

Here's how Meg described the initial stages of her research: The iGEM students, in their brainstorming sessions for the 2006 competition, had decided they wanted to create muscles out of stem cells that could move an object with a contraction. They had one intensive summer to figure out some of the basic mechanisms and get the project off the ground. The team was subsequently divided in two. Each subteam set its sights on differentiating a particular kind of muscle tissue out of stem cells. Shortly after the subteams began their work, a postdoc in a collaborator's lab suggested that, if the team was interested in cell differentiation, they should pursue a project on insulin-producing cells called beta cells, taking a paper that had just been published as their jumping-off point. The process described in the paper was long (about five weeks to go from stem cells to insulin producers) and required a whole host of growth factors, administered in a particular order over the course of cell differentiation. The iGEM leaders subsequently set to work, along with the team, trying to figure out how to devise a simpler beta cell pathway. Meg says:

So the students went and looked at papers, and with our guidance they found some evidence that there are certain cell-fate regulators—cell-fate regulators are the transcription factors that push cells toward a particular fate—that could almost certainly push these cells to make insulin. So if we could somehow turn these cell-fate regulators on in the cell, possibly in a programmed fashion, maybe not, maybe we can get very high efficiency of cells that produce insulin. Because the paper that shows that you just add these cocktails of growth factors comes out with maybe 10 to 15 percent of the cells producing insulin. Not a very high efficiency, but it was a dramatic increase over stem cells that just randomly differentiated. . . . And then after the summer ended, the experiments continued.

In the meantime, the funding for Meg's postdoc had run out and her job was on the line. She remembered the constant specter of unemployment that summer, as she went about her work. One day, she received an e-mail from Ron asking her to put together a presentation on the beta cell work. They had found possible funding. Meg described her lack of preparation and training for the task at hand. Biologists are usually specialized by tissue type or pathway, she explained, whereas the synthetic biologists in Ron's lab were more like jacks-of-all-trades, as long as there was literature to review and access to the necessary instruments and materials.

And so here was a neuroscientist who knew nothing about endocrinology told to become an expert in beta cell biology in five days to speak to experts on beta cells. Just a little pressure. For my job. Because as soon as he said that I knew this was for my job.

The presentation went well, and Meg secured the funding, and her position in the lab. The beta cell research became her primary project. After more journal research, Meg and Ron came up with a plan to differentiate stem cells into beta cells in a two-step process. Meg explained:

So into the literature we go, and we look at what the iGEM students found, we looked up all the different papers involving the differentiation process and found that this process is almost certainly governed by gata factors: gata-4, gata-6, SOX17. . . . So we had plasmids with gata-4 already so I started the process of making endoderm, and it was pretty dramatic. We saw the cells turn from these nice little balls to what's called a cobblestone effect. The cells almost look cuboidal and they're bright green. And it is too cool. It's just a really clear effect.

Unfortunately, the cells were differentiating into the wrong germ layer for pancreatic beta cells. Nevertheless, what Meg had done had been a first, and only quite a while later, she noted, did she come to see the lab's approach as audacious from a "typical stem cell biology" perspective. A year and a half after she had done this work, Meg finally decided to take a course on stem cells:

> And they tell you that you can't directly change stem cells into any of these tissues, and I'm, like, yes you can. No you can't, you have to serum starve them, make them ball up, and then you can differentiate them. And I said, no, you don't have to do that. And they were in total shock. So our ignorance of typical stem cell biology played to our advantage.

Meg's story bears some similarity to James Collins's anecdote. Both pit naïve synthetic biologists against disbelieving experts. And both make the point that experts can be dogmatic, and sometimes wrong.

Both stories also highlight a different feature of ignorance. Ignorance blocks networks. In a piece meant to "challenge the interpretive possibilities of limitlessness" endemic to the network form, Marilyn Strathern turns attention to the technical means of cutting networks, of drawing their boundaries or stopping their flows.[14] Her example of choice is intellectual property among researchers, where "the long network of scientists that was formerly such an aid to knowledge becomes hastily cut. Ownership thereby curtails relations between persons; owners exclude those who do not belong."[15] Since knowledge partakes of circulatory and network idioms, ignorance, too, cuts networks, or prevents them from forming, and curtails relations between persons. Ignorance thus dams the flow of knowledge in ways that can be put to use.

Abstraction versus Detail

Some of the attitude toward ignorance in the lab may have had something to do with broader questions concerning what counts as knowledge of biological systems in the first place. If engineers were heavily reliant on decades of discoveries by those who specialized in these systems, they could also be critical of the way that these other experts assessed and organized what they learned about living things.

When Ron began presenting his work in graduate school and soon after, he found that it was very well received by engineers and computer scientists. It made a splash, and was intuitively interesting to these audiences, even if applications with therapeutic or practical value were still a long way off. Convincing natural scientists of the merit of the work was another story. Referring to the difficulties he encountered with audiences of biologists, Ron explained:

> Especially within a scientific talk, you say all sorts of grandiose things, but then that doesn't make a connection. And at the end of the day they say, well, regardless of what you say, what did you do? What circuit did you make? What proteins did you use? What organism are you doing this in? A lot of times they cannot break away from looking at the details to understand the main concepts. . . . After you do the collection of the data, be able to abstract away from that about what's really going on in the system.

An emphasis on abstraction pervades the engineering approach to synthetic biology. A major part of the engineer's conceptual arsenal, "abstractions" are formal tools for hiding complexity. They are the "parts," "devices," and "systems" whose relationship to each other is construed in terms of a nested hierarchy, with DNA at the very bottom. Ideally, practitioners can focus on different levels of abstraction without recourse to others.

Ron's assessment indicates that the hierarchy is normative as well. It privileges the conceptual and abstract as the levels at which one could gain an understanding of synthetic biological systems. These, argues Ron, were the levels at which meaningful order was to be found in systems designed by engineers. And these were precisely the levels that biologists were missing because of their investment in the "details."[16]

In his comments above, Ron makes a distinction between designed systems and natural ones. Yet Ron's contention that biologists were too bogged down in the "details" to understand some more conceptual or abstract levels of order went beyond how biologists approached engineered systems that had been designed with these abstractions in mind. Ron was more generally critical of the way instruction in biology often emphasized "details" that could themselves, he contended, be better organized according to informational logics. Ron explained,

There's overemphasis on the details and sometimes the big concepts are not emphasized. . . . It shouldn't be like a folklore, kind of passing of information by the fire, using smoke signals and things like that. It should be a much more, I don't want to say rigid, but structured activity of collecting data, putting it into databases, being able to access the data, having it be accessible in machine format so that you can have software that actually looks at the data and assembles it and allows you to make all kinds of claims or hypotheses that you can then test out using computation, using modeling, using abstractions.

In the aftermath of the Human Genome Project (HGP), calls for the reappraisal of biology according to such informational logics have been rampant.[17] These informational assessments trouble distinctions among "knowledge," "information," "data," and the more colloquial "detail," while suggesting that new sources of order and forms of knowledge might perhaps be discovered at greater levels of abstraction. The reassignment of much that is sorted into the bin of "knowledge" into these other bins, like "information" and "detail," produces substantial ambiguity about what amounts to knowledge of biological systems and how such knowledge relates to the activities of engineers.[18]

Complexity and Ignorance

The ambiguity surrounding what constitutes knowledge has left room for the argument to be made that some of the impenetrability and complexity of living things is a consequence of the entrenched knowledge practices of those who have sought to know them. In the early 2000s, the cell biologist Yuri Lazebnik wrote a reflexive essay that argued as much. Not surprisingly, the essay has been popular among those pursuing the engineering approach to synthetic biology. Titled "Can a Biologist Fix a Radio?" the essay argues that biologists, armed with nothing but their methods, would find a broken radio inscrutably complex, whereas engineers could figure out how it works and even how to fix it.[19] Lazebnik's hypothetical radio for the thought experiment is an actually existing radio, one his wife brought back broken from Russia. He uses the radio to parody the investigative methods of biologists and their results, which he sees as, in equal measures, overly local and imprecise. Lazebnik contrasts his parodied biologists with engineers, whose conceptual tool kit, he states, is both more quantitative and more abstract.

Taking aim at his own discipline and its immediate kin, Lazebnik's essay contains the claim that "complexity is a term that is inversely related to the degree of understanding."[20] When Lazebnik suggests that engineers can figure out how the radio works, then, he is not simply arguing that engineers are better at coping with complexity. In Lazebnik's tale, unwieldy complexity is an artifact of disciplinary knowledge practices and methods, which therefore disappears at the hands of engineers.

The notion that complexity might disappear at the hands of engineers resonates with some ideas germinating in recent social scientific scholarship on complexity. As Annemarie Mol and John Law observe in their introduction to the edited volume *Complexities*, across many of the social sciences, a revolt against simplification has been taking place: from history to anthropology, from cultural studies to feminism, "the argument has been that the world is complex and that it shouldn't be tamed too much."[21] In history and political theory, critiques of simplification have focused on rationalization and bureaucracy as the hallmarks of modern state power that reduce complexity by ordering, dividing, simplifying, and excluding. In science and technology studies, Mol and Law identify work that points to problems of scaling up from controlled experimental settings to large-scale technologies or from clinical trials to sick patients. The general shape of these critiques, write the authors, is to argue that "simplifications that reduce a complex reality to whatever it is that fits into a simple scheme tend to 'forget' about the complex, which may mean that the latter is surprising and disturbing when it reappears later on and, in extreme cases, is simply repressed."[22]

Responding to such critiques in anthropology, Hirokazu Miyazaki and Annelise Riles propose an artifactual view of complexity.[23] Miyazaki and Riles argue that complexity may be the result of the failure of analysis. For these authors, much of contemporary anthropology, with its focus on emergence, indeterminacy, and complexity, reflects analytical problems that have fallen out of the account, problems that are hidden from view when complexity is taken to be an inherent feature of social phenomena. The diagnosed complexity of social worlds, for example, may tell us more about the failure of anthropological knowledge than about the sites anthropologists are studying. Social scientific analytical strategies, "in response to the apprehension of the endpoint of their own knowledge, . . . retreat from knowing. And they also retreat from the recognition of the failure of their own knowledge by locating indeterminacy and complexity 'out there.'"[24] In such cases, interpreting complexity as an inherent property of phenomena naturalizes complexity as the cause of failure, rather than its effect.

In her book *Unsimple Truths*, Sandra Mitchell highlights another way in which knowledge practices themselves may give rise to complexity. She notes that complexity sometimes appears between mismatched representations.[25] Arguing that the sciences and public policy often fail to take account of complexity, Mitchell advocates a position she terms "integrative pluralism," which favors "multiple explanations and models at many levels of analysis instead of always expecting a single, bottom-level reductive explanation."[26] Mitchell's argument against reductionism and for the existence of emergent properties (properties that cannot be explained with reference to the properties of their constituents) draws on the partiality of representations. She writes,

> Any representation—be it linguistic, logical, mathematical, visual or physical— stands for something else. To be useful, it cannot include every feature in all the glorious detail of the original, or it is just another full-blown instance of the item it represents. Something must be left out, and what is left out is a joint product of the nature of the representing medium (Perini 2005) and the pragmatic purposes the representation serves. . . . The partiality of any representation leaves open the possibility that the two representations will simplify the phenomena in incompatible ways.[27]

While elsewhere Mitchell seems to locate complexity in the ontology of objects and phenomena, her rejection of reduction turns on epistemic considerations. That is to say, in her account, it is the relationship between representations that sometimes renders reduction impossible. If, then, representations are partial and pragmatic, as she suggests, then complexity may be a consequence of the relationships among representations, or between historically grounded shifts in interests that leave some representations outmoded.

We can tell a story about the last couple of decades in molecular biology that suggests complexity has been increasingly naturalized along these lines. The study of biology has long relied on the simplification of life's complexity, in order to manage and interpret experimental work. A finely tuned middle ground between complexity and simplicity in fact defines the strength of biologists' experimental systems, understood as the "locally manageable, functional units of scientific research."[28] While experimental systems are "machines for reducing complexity" that allow experimenters to exert some control, their power lies in being connected to a network of relations outside of the experimental system that keeps such systems from being trivial or closing in on themselves and enables the generation of anomaly or surprise.[29] A balance between simplicity and complexity is thus built into biology's epistemic practices. Within the experimental systems of biologists, the model organism has perhaps acted as the paradigmatic simplifying tool. Comprising species conducive to certain kinds of research in particular areas and singled out for their relative simplicity, ease of maintenance, and availability, model organisms have produced bodies of research that delineate problems and in turn enable more research.[30]

A concern with managing complexity and its simplification has pervaded the study of biological systems, but nowhere has a commitment to simplicity seemed more foundational and effective than in molecular biology. In the last decade, complexity, therefore, has come as something of a surprise in molecular biology and genetics, particularly in the wake of the HGP. The resurgence of complexity in molecular biology stems partly from decades of qualification with reference to the "central dogma"—famously formulated by Francis Crick in 1957 and summed up with the slogan "DNA makes RNA makes protein"—a model of "precocious simplicity" thought to underlie the relationship between genes and living beings.[31] The central dogma describes a relationship between meaningful information contained in the genetic code, understood as a set of instructions with the "capacity to issue orders," and the implementation of those instructions in the production and maintenance of life.[32] In this account, genes figure prominently.

Yet in recent decades, the central dogma, and its central actor, the gene, have yielded some ground to more "complex" understandings of the relationship between DNA and the life around us. In her book, *The Century of the Gene*, Keller argues that it has become increasingly difficult to pin down what exactly genes are, or, for that matter, what they do.[33] Keller introduces the possibility that genes are on the verge of outliving their usefulness as ways of organizing biological research and thinking about development and heredity.

The HGP played an important role in undermining the centrality of the gene. One of the major outputs of the project was a staggering amount of information. And as sequencing technologies have become cheaper, information has become easier to amass. Eric Lander, a leader in the public effort to sequence the genome, along with Robert Weinberg, wrote in a triumphant piece on the future of the life sciences more than a decade ago, "biology, in the 21st century, is rapidly becoming an information science. Hypotheses will arise as often in silico as in vitro."[34]

The informational terms of genomics have since normalized and become mundane. But what does all this information add up to? The answer is elusive. Thus, the paradoxical result of what has been called "the mapping paradigm," which included massive and in some ways truly effective efforts, was the realization, in the immediate years following the genome sequence's completion, that more information had given geneticists a clearer sense of yawning gaps in knowledge.[35] In the decade leading up to the sequence's completion, the human genome was thought to contain approximately 100,000 genes. By the time the human genome map was announced, the number had shrunk to around 30,000 genes and later 23,000 genes, a figure roughly equivalent to the number of genes in *Caenorhabditis elegans*, the modest roundworm.

For many, the newly revised numbers were a shock, since species complexity had been expected to correlate with numbers of genes. Instead, the relationship between genetic code and the life we see around us has taken researchers to parts of the genome they had been keen on ignoring, and beyond, in what Erika Check Hayden terms a "complexity explosion."[36] They are now haunted by that once-common phrase, "evolutionary junkyard" referring to purportedly noncoding regions.

The image of newly discovered "gaps" in knowledge is a mainstay of post-HGP assessments. The interpretation of what these unknowns amount to follows a glass-half-full (of gaps)/glass-half-empty logic. For some, the emphasis is placed on how little we know and how much remains to be done. For others, the accent is placed on our newly acquired knowledge of how complex living beings really are. The latter, celebratory, stance tends to diagnose complexity as an immanent quality of living beings, a quality so firmly rooted as to be itself available for discovery. The former assessment, on the other hand, tends to emphasize the epistemic side of complexity. Complexity characterizes the gap between our knowledge and some notion of biological systems in themselves.

Indeed, in order to incorporate research discoveries into an understanding of how genomes work, the account has had to increase in complexity significantly. Biological systems are complex, at least partly, insofar as they exceed and undermine the model of gene action around which molecular biology was built. Yet the epistemic side of complexity easily drops out of the account. The result is that the partial failure of a simple biological model is now being naturalized and attributed to the complexity of organic systems. What is more, since the excess is itself produced in relation to a particular way of dividing up the world in molecular terms, it is not entirely clear that molecular biology can be disentangled from the complexity it, at least in part, produced, and in which it is now mired.

These views of complexity have radical implications for the commonly held normative stance that more knowledge is better. This normative stance is self-evident if complexity resides in the phenomena researchers and experts seek to know. However, it ceases to be entirely compelling on the view that complexity sometimes grows out of the theories and models themselves or out of the overlaps, crevices, and cracks between them. In such cases, drawing on additional bodies of knowledge may, in itself, compound complexity. That is, sometimes it might make some sense to simulate the act of starting over.[37]

"*Finito* Experiment and Theory"

What, then, of the relationship between the parts-based approach and biological knowledge? Many observers insist that the parts-based approach to synthetic biology, flush with pragmatic sensibility, is positioned to have limited upshot for our knowledge of biological systems. This view is laid out most clearly in Keller's review of synthetic biology.[38] Keller notes that synthetic biologists of various stripes often invoke a saying Richard Feynman scrawled on his blackboard at Caltech: "What I cannot create, I do not understand." This pithy slogan furnishes synthetic biologists with a sort of motto, pasted as an epigraph or dropped in the introductory lines of research and review articles—ironically, Keller notes, because Feynman has, among scientists, been made to stand for the "ultimately pure scientist."[39]

In Keller's interpretation, synthetic biologists evacuate Feynman's quote of its epistemological significance: "Here, knowing *is* making, and nothing more."[40] Keller quotes one of the most vociferous parts-based approach proselytizers, an engineer by the name of Drew Endy:

> When I go to the hardware store and get a nut and a bolt, . . . I can take those
> two objects, and I can put them together. . . . I don't need to go talk to some

Harvard professor to figure this out. I don't need to do a controlled experiment to see if my first experiment worked. I just get the two objects and put them together.[41]

Keller wryly quips, "*Finito* experiment and theory, enter standardized design and production."[42] For Keller, synthetic biologists' role in "understanding the organization of biological systems as we know them, by Endy's own account, . . . is non-existent."[43] Keller summarizes Endy's purported position: "Synthetic biology's role is not in understanding organisms as they have evolved, but possibly, he adds, in understanding how to remake these organisms to better and more efficiently serve our ends as human users. Here, the aim is to make use of the understanding of biological parts that has emerged from biological science as we have known it to launch an alternative—a synthetic—biology." Keller adds, "But it is not clear what, if anything our expertise in synthetic biology contributes back to an understanding of traditional biology."[44] Yet the invocation of Feynman can be read as precisely questioning the extent to which "traditional biology" constitutes an understanding. We might stop for a moment, then, to appreciate the adoption of Feynman's quote as a motto in all its unabashed chutzpah, in light of which synthetic biologists might have been intent on, *for the first time*, inducing understanding of biology, particularly at a moment when understanding seems so elusive and complexity a source of newly rejuvenated awe.

We can now circle back to the assorted celebrations of naïveté and self-reliance with a more fleshed-out interpretation. The interpretation suggests that synthetic biologists have put the naïve perspective into use as a kind of method, challenging the pithy truism that more knowledge is better. In so doing, they invite us to entertain the possibility that ignorance and amateurism might be more than accidents or matters of convenience in this setting. Ignorance might be a key part of a tool kit that tests the traction of new principles, concepts, and styles of reasoning within deeply colonized epistemic domains, in which knowledge practices and their effects have merged.

Conceding Complexity

As I noted at the beginning of this chapter, wagers are inherently risky things. During my year in Ron's lab, Meg's beta cell differentiation pathway, discussed above, was one of few relatively trial-and-error-free success stories, and it was only a partial success. As I mentioned, her stem cells differentiated into the wrong germ layer. Knowledge of substrate details was therefore often amassed at impasses, through research in scientific journals. In effect, research in molecular biology for these engineers often took the form of "troubleshooting": figuring out what had happened when experiments did not produce anticipated or hoped-for results. More thorough knowledge of biological pathways and processes thus entered through the back door.

Maureen O'Malley has written of the gap between the rhetoric of rational design and the practice of "kludging" in synthetic biology.[45] Kludging is a colloquial term used commonly in electronic and software engineering for a solution characterized by functionality rather than elegance or efficiency. Synthetic biologists have used the term to describe the dominant, and hopefully temporary, problem-solving mode in their field, which involves ad hoc adjustments to particular situations. The term also captures something of the work of remediation when a wager has not born the desired fruit. That is, kludging is an apt metaphor for the process of adjusting a wager and its underlying ethos to the exigencies of the situation. In this case, the dismantling of the barrier between synthetic biology and biological knowledge can be understood as a sort of kludge that slowly increases "complexity" until progress can be made at the cost of generalizability and therefore also at the cost of an engineer's abstract symbolic order.

According to this logic, interdisciplinarity becomes, itself, a sign of the wager's failure. By 2014, a number of high-profile synthetic biologists, Collins among them, concluded that "the most effective practitioners in the further development of the field will be those with an engineering mindset who, functioning alone or in integrated teams, understand most broadly how natural biology works."[46] Ron, too, delivered such a diagnosis in a similar survey of synthetic biology published that same year. "Finally," he explained, "synthetic biology researchers are developing an ever-growing appreciation for biological complexity, which requires interdisciplinary research, new circuit design principles and programming paradigms to overcome barriers such as metabolic load, crosstalk, resource sharing and gene expression noise (and sometimes actually utilize these barriers to create more robust systems)."[47]

In more recent conversations, Ron has explained that those who succeed in synthetic biology have to be "detail oriented," and that "understanding the underlying biology is key." In this sense, over the course of a decade, Ron notes, the field has made a 180-degree turn, though this turn, he explains, also reflects shifting aims for research—whereas many of the proof-of-concept projects, often constructed in bacteria, worked orthogonally to naturally occurring biological systems, many of the ones currently pursued are embedded in cellular processes, which they both sense and modify. Nevertheless, the more recent recognition of biological and cellular complexity marks a shift, and has been linked to the perceived need for more knowledge of the biological substrate through interdisciplinary research.

On Methodological Ignorance

In his introduction to the edited volume *Agnotology*, Robert Proctor critiques the trivializing stance that takes ignorance to be something that always requires rectification.[48] He writes, "Ignorance is most commonly seen . . . as something in need of correction, a kind of natural absence or void where knowledge has yet to spread." Instead, Proctor suggests, "ignorance is more than a void, and not even always a bad thing."[49] Linsey McGoey also warns that much of the work on ignorance tends "to view ignorance as a de facto negative phenomenon, something social actors have an obvious interest in seeking to overcome or to eradicate, and which sociologists have an onus to better identify so that actors are equipped to recognize and combat their own ignorance."[50] McGoey analyzes ignorance from the perspective of actors' interests within organizational settings. From the perspective of the actors she describes, far from being a negative phenomenon needing remedial attention, ignorance proves useful as a tool to combat liability and assert expertise.

In contrast to McGoey's work on strategic ignorance, my line of argument gives ignorance more epistemological weight. It does so by moving from strategy to method, while retaining some of McGoey's emphasis on purposive ignorance. At the same time, the ignorance I have described can be ethically fraught, for it implies ignorance of that which is thought to be known. Thus, taking purposive ignorance as a kind of method involves acknowledging that the line between methodological ignorance and neglect is a thin one, if one exists at all. The troublingly fluid borders among different ethical formulations of ignorance are grounds for taking interest in methodological ignorance and the problems it raises, not for ignoring it.

I therefore conclude this chapter by recalling an incitement to ignorance, laid out with great panache in the methodological anarchism of the philosopher of science Paul Feyerabend.[51] Arguing that "prejudices are found by contrast, not by analysis," Feyerabend suggests that the different pieces necessary to test a new theory or conceptual system could not possibly all show up on the scene at the same time.[52] This is in part because observation, including the trivial-seeming kind, carries a kernel of naturalized interpretation within it, the latent signs of a bygone era or the theory-laden tectonic marks of a prevailing view. An entire apparatus of sciences and instruments is required to bring observation into alignment with a new theory or conceptual system, but these often lag. The way to proceed, as Feyerabend famously puts it many times, is counterinductively. A new conceptual system is thus not tested by rational procedure and compatibility with observation. Instead, Feyerabend writes, the test of a conceptual system comes after we "*wait* and . . . *ignore* large masses of critical observations and measurements."[53]

Not all ignorance is good, for Feyerabend. He laments the ignorance of anthropologists who failed to understand, much less record, the astronomical and cosmological beliefs of different groups, for example. Ignorance is by no means a diagnostic of good practice but part of a plural world of methodological tactics tidily excluded from view.[54]

Feyerabend's incitement to ignore is illuminating in another way. In ignorance's tie with waiting, ignorance becomes nail-biting work. It requires a carefully blinkered view and the ability to bide one's time, something that engineers, many of whom tethered their funding to technological promises, didn't necessarily build into their research programs.

Chapter 3

Looking for Patterns

In chapter 2, I analyzed an epistemological wager at the heart of the parts-based approach to synthetic biology and described its unwinding in the face of a mixed experimental track record. One thing to note is that this mixed track record is not calculated by averaging clear-cut successes and equally clear-cut failures, because synthetic biological experiments rarely produce such neatly categorizable results. Despite all the laudatory attention paid to synthetic life-forms that conform to their designers' intentions, and the critical attention paid to the projects that do not, many projects wind up somewhere in between success and failure: they work more or less. This middle territory is where most of the experiments I observed spent a significant amount of time. In the context of experiments with modified life-forms, ambiguous experimental results take the form of a vast array of biotic not-quites: somewhat rogue, somewhat unruly, these organismic by-products point in various directions when it comes to figuring out how much control synthetic biologists have over their designs and what steps should be taken as correctives. In this chapter, I focus on two episodes, one from each lab, hewing my attention more closely to experimental processes and tracking the ways that practitioners tackle, reason, and think through puzzling experimental results in their attempts to move forward and to figure out what's next.

In the first episode, in mid-October 2010, Sarah, a rising senior, and Betsy, the new postdoc—both in Michael's lab—present findings at a lab meeting. Sarah and Betsy are reporting on the latest round of experiments, and their tone is markedly tense. They had both recently taken up the synthetic biology project, following the departure of a very talented undergraduate and a couple of graduate students who had been pursuing the work, all of whom had recently graduated. Betsy is taking a leadership role in the lab, having arrived at Princeton a few months earlier, after the completion of her doctoral thesis in chemistry at Yale. Sarah, a hardworking, moderately shy premedical student, who reports to both Betsy and Michael, is contributing to the project as part of her senior thesis, a required element of Princeton's undergraduate curriculum.

The problem on this day is that this new team is having trouble replicating the results of experiments that had succeeded in the past for students who are long gone. A key control is producing odd results with a number of different bacterial strains. Recall that Michael's lab's main project tests the function of synthetic de novo proteins in barely viable bacteria called auxotrophs. Auxotrophs are mutated E. coli strains that can grow when fed a nutrient-rich diet, but cannot grow on poor media. Michael's lab had already tested many of their own proteins for their ability to compensate for the deleted functions in the auxotrophs. They had managed to "rescue" four E. coli strains, demonstrating that novel proteins can support life.

For the auxotrophs project to work, one set of controls has to show that these strains are in fact auxotrophs: cells that rely on their environment for nutrients that their naturally occurring and unaltered bacterial counterparts can produce. But several of the auxotrophs show what may be some growth on minimal media; rather than exhibiting the trademark traits of E. coli colonies, however, petri dishes containing these particular E. coli strains are covered in tiny grayish dots. The mysterious specks are more than mere inconvenience. Their presence possibly undermines an essential part of the study design meant to prove that synthetic genes are solely responsible for growth, threatening a well-ordered conceptual and material network of experimental relations—that is, if these specks are indeed cell colonies.

Back in the lab meeting, Sarah and Betsy come up with a series of hypotheses as to why the phantom colonies might be showing up and explain how they experimentally knocked these alternative explanations out one by one, as best they could. Sarah explains that they have become incredibly meticulous about the way they run experiments in order to rule out human error and contamination. She now sees colonies everywhere, she laments, so that she no longer trusts her own perception and asks other lab members to look at her plates. Michael asks her lightheartedly if she dreams about cell colonies too, attempting to lower the anxiety level of a very concerned undergraduate unsure of her own hand in what seems for the moment like an experimental unraveling.

Twenty minutes into the lab meeting, a question that had begun to take shape implicitly is finally made explicit when Betsy asks, "How do we define growth?"[1] Years of research in the auxotrophs project, in which practitioners passed judgment on growth and its absence on a daily basis, had gone by without the notion of growth ever having needed to be specified. A debate ensues. Lab members disagree on whether to call the irregular grayish dots growth. An official error margin for "no growth" is suggested: Dead . . . +/−5. Defiantly, the most senior postdoc interjects that if the phantom dots do not expand, they should just be ignored and tallied as no growth. This solution is quickly and vociferously squashed as scientifically suspect, affirming the notion that labs are local moral universes. Plus, the phantom colonies have caused too much unease, and occasioned too much work, to just be ignored. The collective sentiment regarding this solution is somewhat reminiscent of the famous words of Miracle Max in the movie *The Princess Bride*: "There's a big difference between mostly dead and all dead."[2]

I sat down with Sarah and Betsy a few times in the months that followed. They each insisted that the questions surrounding minimal cell growth arose because they were new to the auxotrophs experiments, though, listening to them, I soon realized that their understanding of the effects of their newness followed two separate logics. First, there was the way changes of personnel affected the conventionalization and routinization of laboratory practice. Betsy explained:

So, Sarah started shortly before I got here. I got here in late March. She hadn't been here that long. I got here and we had only a slight overlap with another undergrad who had worked on the project and no overlap with any grad student. If there's anyone to overlap with, it's [that undergrad], but you know, [the undergraduates] are not around as much. Especially her, she was finishing up her thesis, she would do stuff at night, so there wasn't really a lot of overlap with her. So (a), Sarah and I figured it's probably human error. It's probably our fault, right? And (b), if, say, John [the previous graduate student on the project] had still been here, he would have been, like, oh, that's normal, seen that with six other strains, don't freak out. But I think we freaked out a little bit because we were new to the project, we were new to the lab.

Experiments often come packaged together with the choices, standards, and observational skills of others. Some such packaging is explicit. So, for example, the auxotroph experiments were all conducted at 37 degrees Celsius because of some rather incidental research a graduate student had done on toxicity some years earlier. The norm for these bacteria would be 34 degrees Celsius. Betsy speculated that this decision had been pretty arbitrary, but everyone with their hands in the auxotrophs research was now stuck with it. And anyway, as she explained, their growth conditions were themselves a matter of some arbitrariness. They could decide to observe growth for three days, in which case neither the phantom colonies nor many of their other more robust results would have turned up, or they could wait thirty days and observe a lot more growth than they had. Betsy and Sarah were both fairly confident that their predecessors had seen these grayish specks before, and had made some decisions about how to treat them, backed by some rationale. To know how they viewed them would suffice for upholding a consistent communal standard across practitioners, time, and space, even if the precise state of these cells remained opaque. In this way, skirting ambiguity could be a matter of relying on settled laboratory convention rather than careful adjudication of the meaning of growth.

Sarah saw another issue with her newness to the project. She chalked up many of the difficulties she encountered to her own inexperience with cells. Lorraine Daston has drawn attention to the ontology of scientific observation, that is, how experts stabilize the observation of particular things.[3] Criticizing the hold of what she terms "filter" views, which place their emphasis on the contaminating subject's precarious contact with the objective world, Daston invites scholars to pay attention to the temporal process through which experts move from initial perceptual disorder to order: "The novice sees only blurs and blobs under the microscope; experience and training are required in order to make sense of this visual chaos, in order to be able to see *things*."[4] I return to Daston's intervention later in this chapter; for now, we note that Sarah's understanding of her own trajectory traces such a path. In gaining experience, Sarah had been able to extend the range and stability of her perception and, in the process, had become more independent in her work. She said:

> I didn't know if what I was seeing were crystals or colonies or fake colonies. . . . And I wasn't yet accustomed to looking at so many different types of colonies. Now I can tell if things are smaller than normal, or not the right color, or they're growing in a weird amount of time, but back then I didn't know what to compare it to.

While Sarah traces an increasingly ordered set of perceptions, these perceptions reflect gaining skills in the local universe of observations that hold for the experiments with which she works. In Sarah's case, "smaller than normal" or "not the right color" are relative terms that reflect familiarity with the experimental system as semi-self-enclosed. It is in relation to the system itself that deviations are taken to be meaningful. Such relative observations may be part of the training process, a stopping point between novice and expert. But Sarah's relative statements also reflect an exigency of the research in which she is engaged, defined as it is by making new experimental things and generalizable to many settings in which researchers produce experimental aberrations, in synthetic biology and beyond. Normal and abnormal *are* in some important sense local coordinates, insofar as Michael's lab makes organisms that do not exist in nature, or in other labs. In this sense, we can speak of *local synthetic biologies*, which array the normal and the pathological around a particular technoscientific laboratory practice bounded by the singularity of the biotic things it produces.[5]

Contrast Sarah's description of how she interprets cells with that of Meg, the postdoc in Ron's lab who had been culturing cells for many years and had extensive experience with biological experiments. When Meg described the results of some recent experiments to me, she gave voice to the fullness of the observational repertoire she used to identify cells and the variety of their possible forms. After Meg had managed to differentiate stem cells beyond the endoderm phase, she described her resulting cells as "big, ugly colonies." These colonies were very hard to image on a microscope, because they looked like "a glob of death." Yet she knew they were alive. I asked her how she knew this. She answered,

> Well there are several ways. You can actually use propidium iodide to look, and that stains live cells. Or you can just know. If you've looked at cells long enough you can tell what a dead cell is versus what a live cell looks like. And live cells are very three-dimensional, you'll see shadowing and you'll see the color throughout the cytoplasm, whereas if they're dead you'll see punctuated color, because that's chromosomes breaking, and breaking chromosomes have their own autofluorescence, so that complicates things, because you still see red and green but you have to look closer at the cells and really make sure that you're seeing the fluorescence throughout.

If Sarah's relative statements lean on the "normal" appearance of results, Meg's description points beyond the bounds of her own experiments. Moreover, Meg's insistence that "you can just know" might capture something of the ineffable quality of experiential knowledge, its tacit dimensions, or it might instead reflect the instantaneousness—what Daston calls the "all-at-onceness"—of the perception and judgment of a well-trained observer.[6]

So how did Sarah and Betsy resolve the questions surrounding what counts as growth? In the course of mulling over the controls that showed what could be slight growth, Michael, Betsy, and Sarah had come up with a new category of experimental results. The claim with reference to the *E. coli* strains in this category was significantly weaker, but still worth making, they decided. The problem of the definition of growth was therefore deferred through the delineation of a category for the indeterminate results. The indeterminate results were thus sequestered from the more robust results of other experiments while also opening up a new axis of distinction in the experimental system as a whole. Betsy explained:

> So we did also determine that there were a few strains that were growing after a really long time, like eight or nine days. So, there was one that Sarah was working on that she determined that, if it grows with no additional proteins in eight days, but we can get it to grow in two days with the additional protein, that's still a phenotype. It's a slightly different phenotype than we were originally looking for, which was a rescue of a total auxotroph, but we did sit down with Michael and talk about whether that was something worth pur-

suing, and we decided it was. Obviously, we have to be careful that we don't imply that it's rescuing a totally dead cell. . . . It's more like it's making them healthier, as opposed to rescuing from the dead.

The introduction of the new phenotype incorporates an interest in the relative health of the cells. It recognizes the phantom cells as probably living but certainly not thriving. All these phantom cells have to be is significantly worse off than the alternative with the synthetic protein. But the introduction of the new phenotype, Betsy explained, involves a trade-off: sometimes, rather than complicate the claims, it's better to just move on.

Turing Patterns

I draw the second episode from Ron's lab. It's the end of June. The iGEM team formally began its tenure in Ron's lab three weeks ago, but team members have yet to settle on a project. Ron stands at the front of a conference room in the electrical engineering department (site of at least twice-daily lab meetings during these busy summer months), projector fired up, ready to give a PowerPoint presentation of his own. I take a seat at the long oval conference table alongside the dozen or so team members, and the graduate students and postdoc leading the team. Ron begins his presentation, aimed, I quickly learn, at selling the team on a possible project.

The project Ron has in mind, and on which some graduate students had already done some preliminary work, involves Turing patterns, an area of interest for both biologists and engineers. Toward the end of his tragically foreshortened life, Alan Turing turned his attention to biology and chemistry. The problem he sought to interrogate was the development of patterns: in particular, how initially uniform bits assume complex forms and functions that display nonuniformity. In biology, those bits are often cells that go on to produce patterns like those found in fur and on fish scales. The now classic paper in which Turing offered a theoretical account of these patterns, titled "The Chemical Basis of Morphogenesis," was published in 1952.[7] Turing's reaction-diffusion model relies on two chemical species, both of which diffuse. One of the chemicals, termed the activator, activates its own synthesis, while also activating synthesis of the second chemical species, called an inhibitor, which inhibits synthesis of the activator. The key is that the two diffuse at different rates, causing an initially homogeneous distribution to produce patterns. Change the parameters of the model (i.e., the rates of diffusion and decay), and you get a diversity of patterns.

Back to the presentation. After Ron's first slide introduces Turing patterns in very general terms, the second slide provides examples of purported Turing patterns in nature. These are *National Geographic*–style images of animals in their ostensibly natural habitats: bright tropical fish gleaming in water, a cheetah mid-leap, and a profile view of a languid giraffe. Ron draws the audience's attention to the patterns in scales and fur, pausing for a moment for those seated in the room to appreciate the photos. The pictures of the animals, at this point in the project's life cycle, are accessories. They are not pitched as objects of thought but rather as illustrations of the kinds of naturally occurring patterns for which Turing attempted to account. They also serve as a bridge between a set of technical-seeming equations and the kinds of phenomena these equations could yield at the hands of synthetic biologists. Nonetheless, when Ron finishes, the students, slumped in their chairs, are not sufficiently drawn in, and so they pass on the idea. Instead, Wei, a postdoc in the lab who had already made some of the genetic constructs for the Turing patterns, picks it up.

Over the course of subsequent months, I kept track of the Turing patterns project as it went through several iterations. After several rounds of tweaking experiments and models, Ron and Wei would confront the realization that what counts as a pattern is anything but self-evident. Petri dishes smattered with fluorescent cell colonies growing in irregular shapes would be the undoing of the self-evidence of patterns.

And then there were the animals whose images had been recruited as benign exemplars of patterns. Their role too would change. They would become objects of thought and furnish an unanticipated source of wiggle room in a network of epistemic, material, and rhetorical relations derailed by hard-to-interpret results. Tethered as they appear to be to reality, even the animals cannot escape some revision when experiments produce ambiguous results.

In *Toward a History of Epistemic Things*, Hans-Jörg Rheinberger discusses the necessity for conceptual indeterminacy in experimental work.[8] He begins with Freud's opening remarks on science and concepts in "Instincts and Their Vicissitudes."[9] "What Freud invites us to ponder here," writes Rheinberger, "is the ineffable trace of scientific action or, as I would like to call it, the *experimental situation*. It appears as if the relationship of 'deriving' ideas from the material of observation and of 'imposing' ideas upon that material represented the focal point of the argument."[10] The "deriving from" and "imposing upon" relationship to which Rheinberger refers gradually articulates the basic concepts of a research program. This movement goes hand in hand with the duality of concepts, which are, at once, instruments of experimental investigation and the products of such investigation. The aim of all this simultaneous movement between loosely defined concepts and observation for Freud is not to settle on definitions, but rather to render the "sensed" explicit, and thereby make room for new observations, explains Rheinberger.

The progress of the Turing patterns project illustrates some similar dynamics involving imposition and derivation surrounding "patterns." Yet Rheinberger's characterization of the relationship between concepts and experimental systems starts in the middle of the life cycle of concepts, while Freud directs us also to their beginning. Freud reminds us that the concepts that are imposed on observations must come from somewhere other than the material of observation at hand. That is, initial formulations of concepts may have nonexperimental lives as well: they originate in prior conventional usage.[11]

In pursuing this supplement to Rheinberger's argument, I return to Daston's concern with the ontology of scientific observation, this time with an important caveat regarding the primordial state from which "things" emerge. Here, I provide an account of how an engineer learns to see biological things that pushes against the notion of the naïve or innocent eye (a theme I pick up from chapter 2) that accompanies a certain understanding of the "chaotic" starting point. The interpretation I provide therefore incorporates some of the gestalt elements that Daston seeks to leave behind. Where Daston sees the ontology of scientific observation as a learning process, I suggest that it might still be fruitful to think of it as, in some cases, a process more akin to conversion.

Though Turing's classic work on morphogenesis was first published in the 1950s, Turing's reaction-diffusion system didn't gain traction among biologists until the 1970s, and even then, a short flurry of writings in favor of the notion that Turing's model did indeed underlie diverse natural patterns was followed by much argument in favor of a different explanation. In the late 1960s, a British developmental biologist by the name of Lewis Wolpert had proposed the theory of *positional information* to account for multicellular patterns. According to Wolpert's theory, the concentration of a diffusible molecule varies by location within a concentration gradient. By sensing concentration, cells can locate themselves spatially in the field and differentiate into different cell phenotypes.[12]

The turning point in favor of Turing's version of morphogenesis came in the 1980s, when chemists synthesized Turing patterns in the lab, carefully controlling reactions and producing computer simulations. Since then, biologists have continued to uncover evidence in support of Turing's model. By forward engineering simple circuits, synthetic biologists contend that their approach could offer a way of clarifying how Turing patterns might work in biological systems. From a technological standpoint, the ability to pattern growth without having to place every element or genetically engineer many irregular bits would come in handy for many of the kinds of technological systems synthetic biologists are interested in building.

If the reader will forgive a brief anthropological digression at this point, there is perhaps another story we can tell about the significance of cellular patterns for synthetic biology. In a discussion of the development and rationalization of Prussian forestry in *Seeing Like a State*, James Scott makes a fascinating, if fleeting, conjecture about the geometric patterns in which forests were planted.[13] In Scott's book, the development of early forest management serves as a paradigm case for the unintended consequences of modernist order. In the midst of his account of Prussian forestry's eventual failure, Scott ventures an answer to a seemingly trivial question: Why were the forests planted in neat rows? Contrary to the assumption that spatial order is rationally justified, Scott suggests that rows are nothing more (or less) than the aesthetic manifestation of the ideal of planning. That is, geometric regularity accompanies planning not as a necessary outgrowth, but as a *symbol* of domination through rationalization itself.

Of course, geometric order has never been the exclusive provenance of the heirs to modernity. The predilection for geometric forms among "nonmoderns" was a problem that occupied no less an anthropological luminary than Franz Boas, who found the prevalence of these forms among the tribes he studied hard to account for. The issue, for Boas, was that geometrical elements were not, by the assessment of his own observational skills, common in the natural world. Of geometrical forms, he wrote, "They are of such rare occurrence in nature, so rare indeed that they had hardly ever a chance to impress themselves upon the mind."[14] It was the famed Austrian art historian Ernst Gombrich (whose well-known work on the philosophy and psychology of perception in the visual arts I return to later in this chapter) who commented on the peculiarity of Boas's bewilderment, considering that Boas was "so much aware of man's enjoyment of mastery."[15] In his treatise on the decorative arts, *The Sense of Order*, Gombrich ventured a hypothesis aimed at explaining the pervasiveness of these patterns, when he wrote, "The conclusion to which we are driven suggests that it is precisely because these forms are rare in nature that the human mind has chosen those manifestations of regularity which are recognizably a product of a controlling mind and thus stand out against the random medley of nature."[16] For Gombrich, too, geometrical forms, as such, are symbols of control.

Wei and Ron went about designing the genetic circuits that would hopefully yield patterned growth in *E. coli*. Here's how it worked: Using eight genes regulated by five promoters, the genetic circuits were engineered to secrete two artificial diffusible morphogens. The first morphogen, 3OC12HSL, activates its own synthesis, as well as the synthesis of the second morphogen, C4HSL. C4HSL, in turn, inhibits the synthesis of both morphogens. To break symmetry and create patterns, Turing's reaction-diffusion system requires that the two morphogens diffuse at different rates. Toward that end, in Ron and Wei's setup, 3OC12HSL diffuses more slowly than C4HSL. By varying the concentration of a small-molecule inducer, experimenters can determine what pattern the system will make. And patterns are made visible by the inclusion in the system of genes that control the synthesis of red and green fluorescent proteins. Each color's expression is correlated with high concentrations of one of the two morphogens.

The experimental setup was accompanied by a deterministic reaction-diffusion model for the lab's genetic constructs. The model explicitly incorporates descriptions of various chemical reactions in the system, including synthesis and diffusion of the morphogens. Visually, the deterministic model produced neat and tidy honeycomb patterns.

After designing and modeling the system, Wei went about running the experiments. He genetically engineered the cells to contain the reaction-diffusion circuit design, then cultured the cells in liquid media, plating them on a petri dish to form a homogeneous distribution with low population density. The petri dishes were then incubated to encourage growth.

The results were less than stellar. On the one hand, the experiments did outperform a key control meant to show that the patterns were generated by the engineered cells sensing and responding to each other rather than through spontaneous, independent cellular processes. On the other hand, the glaring discrepancy between the model and the experiments was immediately apparent. The model produced neat shapes, evenly distributed in space, whereas the distribution of fluorescence in the experiments looked splotchy and irregular, a smattering of variously sized and shaped blotches. These were not the "patterns" Wei and Ron were hoping for. They now had two images in play: the honeycomb of the model, and the blotches of the genetically engineered cells in the petri dish. What next?

Ron explained to me that, among a design goal, a model, and an experiment, an elaborate negotiation takes place. Ideally, the model and the experiment produce the same results right off the bat. This, he lamented, never happens. The model and the experiment are therefore herded toward a point of convergence, and the choice of which one must be made to accommodate the other is itself not trivial. Ron explained, "So there's a goal, and then how quickly can the model get to the goal, and how quickly can the experiment get to the same goal, and how do they converge." That is, not just any convergence will do. The engineering goal delineates where the two should converge. Successive iterations may bring the experiment and the model into greater agreement while the two slowly stray away from the engineering feat to be accomplished. In the case of the Turing patterns, while Ron began with a sense of the self-evidence of patterns, the difficulty of pinning down a convergence point undermined this self-evidence. In our conversations, Ron came to ask, "What is a pattern?" and to frame this rather broad quandary as "the million-dollar question." The entire experimental apparatus had been built around patterns. That same experimental apparatus now raised questions about the terms in which it had been forged, showing them to be vague and underdetermined.

In the case of the Turing patterns project, it was the design goal itself that proved fuzzy. The indeterminacy of the design runs against some fairly well-lodged theoretical impulses in the study of creativity. In their introduction to their edited volume *Creativity and Cultural Improvisation*,

Tim Ingold and Elizabeth Hallam make a case for approaching creativity from the side of improvisation, rather than innovation.[17] To explain the difference, they recruit a paradigmatic opposition: the architect versus the builder. Building, as it were, on the writings of Stewart Brand, they explain,

> A famous architect designs a building, the like of which the world has never seen before. He is celebrated for his creativity. Yet his design will get no further than the drawing board or portfolio until the builders step in to implement it. Building is not straightforward. It takes time, during which the world will not stop still: when the work is complete the building will stand in an environment that could not have been envisioned when it started. It takes materials, which have properties of their own and are not predisposed to fall into the shapes and configurations required of them. And it takes people, who have to make the most of their own skill and experience in order to cajole the materials into doing what the architect wants. In order to accommodate the inflexible design to the realities of a fickle and inconstant world, builders have to improvise all the way. There is a kink, as Stewart Brand writes, between the world and the architect's idea of it: "the idea is crystalline, the fact fluid" (Brand 1994): 2). Builders inhabit that kink.[18]

In the passage above, the design is a fixed coordinate in ideational space, as opposed to the building itself, which will have to be set in an inconstant environment and made of some fickle substance.[19] For the moment, let's accept the Platonic premise that there is something peculiarly fixed about the architect's idea. Most designs take shape in some mediating form before they get built or assembled, and these forms possess their own exigencies and affordances. In this sense, whether or not "the idea is crystalline," the method of rendering the design, and of conveying it to the builder—be it language, sketch, plan, or blueprint—already introduces enough perturbation to suggest that designing and building are conjoined and that design, as a process, is more improvisatory than crystalline. Or, rather, design involves a lot of "building" too. However, we can question the original assumption that the design has some crystalline preexistence in some ideational realm. Designs are also underdetermined and flexible, and therefore open to some improvisation. Such improvisation can be understood not as a way of going beyond the design's fuzzy boundaries, but rather of exploring the space within them.

Ron and Wei had been working on the Turing patterns project for about nine months when I saw Ron present the results, this time in his graduate seminar on synthetic biology. Recall that, in the initial presentation, the animal images served as examples, rather than food for thought. Their role was auxiliary to the substantive architecture of the presentation. They were there to inspire curiosity and make a connection to a real-world phenomenon.

This time, much like in the initial presentation to the iGEM students, Ron commences with the pictures of animals, coats shining in the sun, scales glistening underwater. Ron then explains the genetic constructs, the experiments, the model, and the results, and ends with an intriguing coda. After showing the perfect patterns of the model, and the blotchy semi-patterns of the experiments, Ron goes back to the slide with the animals. He zooms in on the giraffe and points to the irregular domains and spots. "If you look closely at the giraffe," Ron points out, "the domains vary and the pattern is not uniform." Having produced a distribution of cells that does not accurately reproduce the perfect circles and spacing of the model, Ron goes back to the biological occurrence that served as the project's initial inspiration. And now, rather than tinker with the experiments, the model, or the design, Ron tinkers with the visual perception of the giraffe. That is, the giraffe is reexamined in light of the experiments. Ron points out the irregularities in the giraffe's fur, the asymmetries, the heterogeneity in shapes and colors. The giraffe now seems to be on the verge of losing its patterns.

The reexamination of the giraffe affects what happens next. Ron uses the giraffe to argue that the experiments' messiness is a better approximation than the model for what Turing patterns, executed in a biological substrate, should look like. In this way, the giraffe bolsters the decision to tinker with the model first, rather than the experiments.

A new interpretation of the natural example thus helps arbitrate differences between the model and the experiment, and the experiment tentatively wins. The decision resonates with Ian Hacking's observation that "our preserved theories and the world fit together so snugly less because we have found out how the world is than because we have tailored each to the other."[20] In this case, we see how "the world" is subject to reinterpretation in light of experimental results. The shift exemplifies the tentative place of perception. It can be made to play for different teams. And so, Ron and Wei introduce a stochastic model that inches closer to the experiments.

Seeing Patterns

When I first presented this episode from Ron's lab to an interdisciplinary audience, an incredulous audience member asked me, "Did Ron really not see that giraffes are messy when he started?" Obviously, I couldn't answer his question. But I conjectured that Ron simply wasn't looking for mess. If someone had said to him, "Wow, aren't giraffes messy," maybe he would have agreed. It might have even sped things along a little bit. The audience member wasn't satisfied. He explained that, from an evolutionary standpoint, a perfectly patterned giraffe makes no sense. I was struck by the appeal to theory, rather than unbiased perception, to substantiate the point about giraffe patterns. The spectator's incredulity was grounded not in the obviousness of messy patterns to the naked eye, but rather in the obviousness of the evolutionary standpoint and its related perceptual schemas, which render neat patterns unlikely, if not impossible.[21] In this sense, the problem could be construed not merely as a perceptual one, but, rather, as a perceptual one mediated by theory.

In her framing of the ontology of scientific observation referenced earlier in this chapter, Daston argues that the philosophy of science is plush with neo-Kantian "filter" metaphors that suggest a particularly hermetic subject who imposes cognitive constraints on an equally hermetic reality. For Daston, the philosophers of science Norwood Russell Hanson and Thomas Kuhn serve as good representatives of the "filter" view: Hanson, through his analysis of theory-laden observation;[22] Kuhn, through his reliance on gestalt psychology to explain the relationships among scientific paradigms.[23] Daston contrasts such views with those of the Polish bacteriologist Ludwik Fleck.[24] Though readings of Fleck often assimilate his insights into the Kuhnian paradigm framework, Daston follows Bruno Latour in noting that such readings downplay Fleck's emphasis on the work of both experience and time. Daston summarizes Fleck's position as follows:

> For Fleck, learning to see like a scientist was a matter of accumulated experience—not only of an individual but of a well-trained collective. The fault

line in epistemology did not run between subjects and objects, the great Kantian divide, but, rather, between inexperience and experience. Unlike the neo-Kantians, who worried about how the subjective mind could know the objective world, Fleck was concerned with how perception forged stable kinds out of confused sensations. For the neo-Kantians, the problem was the gulf between the subjective and the objective; for Fleck, it was generating order out of chaos.[25]

Ironically, then, in Ron's case, it was a surfeit of order that furnished the starting point. This is because Ron did see "things" at the outset. He saw patterns. His was not the naïve observation of the complete novice, at sea in chaos. His was the stylized observation of someone who approached a giraffe, some petri dishes, and a model with a parsimonious, and beloved, set of equations in hand, namely, Turing's.

In his classic work on art and perception, Gombrich, himself a superlative representative of the "filter" view in the arts, catalogs pictorial misrepresentations of animals through the ages: an Egyptian limestone relief depicting plants the Pharaoh Thutmose brought back from the Syrian campaign said to depict "the truth," yet unrecognizable to a modern-day botanist; a drawing by the medieval gothic builder, Villard de Honnecourt, of a "curiously stiff lion, seen *en face*," entirely ornamental in appearance but captioned "*Et sacies bien qu'il fu contrefais al vif*" ("know well that it is drawn from life"); a letterpress of a German woodcut from the sixteenth century depicting "the exact counterfeit" of a locust whose gait is unmistakably horse-like.[26]

Among his examples, Gombrich includes a Roman engraving dating from 1601 depicting a whale washed ashore near Ancona. The whale engraving, Gombrich shows, bears an uncanny resemblance to an earlier print drawn by a Dutch artist in 1598, of which it is most likely a copy. "But surely," muses Gombrich, "the Dutch artists of the late sixteenth century, those masters of realism, would be able to portray a whale? Not quite, it seems, for the creature looks suspiciously as if it had ears, and whales with ears, I am assured on higher authority, do not exist."[27] Gombrich conjectures that the Dutch artist "mistook one of the whale's flippers for an ear, and therefore placed it far too close to the eye. He, too, was misled by a familiar schema, namely the schema of the typical head."[28] Gombrich surmises that these images all exhibit the particular tendency to classify the unfamiliar with the familiar: "Without some starting point, some initial schema, we could never get hold of the flux of experience. Without categories, we could not sort out impressions."[29]

The implication of these various images for Gombrich's work is that the phrase "the language of art" is more than an apt metaphor. Artistic practice is made up of conventional signs composed of broadly accepted schemas, rather than natural signs.[30] To recognize that art is made up of conventional signs is to open the door for the very possibility of a recognizable artistic *style*, the problem that motivates Gombrich's study in the first place. The correct portrait, explains Gombrich, in finally taking on the problem of the relativism of perception, "is not a faithful record of a visual experience, but the faithful construction of a relational model."[31] One can never merely copy nature or reality: "The innocent eye is a myth."[32] For Gombrich, as for the philosopher Nelson Goodman, who found in Gombrich's work material for his own conventionalist view of art, primordial chaos is never that primordial. It is always already styled. Both Gombrich and Goodman echo the Kantian dictum that "the innocent eye is blind and the virgin mind is empty."[33] Under such a reading, Ron's messy giraffe is a matter not of learning to see giraffes in all their messiness, but of conversion from one perceptual style to another. Moreover, since that conversion is guided not by evolutionary theory but by synthetic biological experiments, the unity of the messy terminal point reached through these two different routes is itself an open question.

The giraffes never made it into the draft paper. They were tools of thought and persuasion that were recruited in more informal settings, then discarded like scaffolding in formal ones. And while the Turing project didn't produce publications in the year in which I tracked its progress, it did bolster already simmering interest in the lab in understanding the effects of "noise" on biological systems, illustrating that while ambiguous results may take a long time to generate publications, if they ever do, they may nonetheless guide attention to new problems and redraw the contours of perception and observation.

Chapter 4

To the Editor

Synthetic biologists who pursue an academic path must contend with the grand arbiter of academic success: peer review. When I began visiting Michael's lab regularly, a key paper the lab was attempting to publish had just been rejected from a highly respected peer-reviewed journal devoted to noteworthy and general scientific work. Michael was miffed. As is often the case in journal publishing, the editor of the journal had solicited three anonymous reviews in the initial round and returned the reviews to Michael along with the rejection letter. Michael felt that the reviewers had misunderstood the work, so he and his coauthors in the lab subsequently set out to craft a "rebuttal" to the reviews, to be sent to the editor, an uncommon move after a paper has already been rejected, though not unheard of. Michael then sent me all the documents pertinent to the exchange, mostly for posterity, but also, I suspect, to recruit a witness for what he saw as scholarly injustice. From an anthropological perspective, the rebuttal constitutes a rare documentary artifact of modern knowledge practices.

At face value, the rebuttal provides some insight into the ways authors read and interpret peer reviews, but this insight must be filtered through some general communicative constraints. Academics in many fields—chemistry and synthetic biology among them—write for different readers and adopt different styles and genres.[1] There are the papers, review articles, and grant applications; the peer reviews, tenure letters, recommendation letters, and e-mails; all of which are fairly practiced affairs, especially for an old hand like Michael. A rebuttal, while perhaps less thoroughly scripted because the crafting of such a document is a rarer occurrence, is scripted nonetheless. The limits of academic civility to some extent shape and circumscribe what gets said, and how. Moreover, the rebuttal is a strategic document aimed at convincing a journal editor to change their assessment of the paper and thus engages the reviews in various ways meant to appeal to the editor's higher authority. The document therefore moves among distinct tactics and moods, registering grievances, admitting fault, and debating factual claims, all the while insisting on the merit of the research.

At the center of the exchange among Hecht et al. (the author function in question), the anonymous reviewers (sometimes termed referees), and the journal are two linked problems. First, Hecht et al. read the referee reports as raising questions about the trustworthiness not merely of the lab's results, but also of the lab. The rebuttal is therefore as concerned with virtue as it is with knowledge claims, though the two are perhaps not as easily separated as one might assume. Against the polemical insistence that we live today in a "postvirtuous" society, Steven Shapin has argued that technoscience remains a moral vocation.[2] The exchange with the journal offers fertile ground for thinking about the textual dimensions of the moral bonds that bind communities of experts. Initially buried in the impersonality and subjectlessness of the passive-voiced draft paper, the lab-as-moral-collective is subsequently excavated from between the paper's lines, revealing, in the process, the inherent tension between moral presence and subjective absence that has characterized much modern science, and that carries into technoscientific settings.[3]

The second problem involves the main concern of all three referees: Hecht et al. do not provide an account of *how* their de novo proteins rescue the auxotrophic bacteria. The hesitation about the validity of the research absent a causal account hinges on the possibility that, since Hecht et al. do not know how their proteins work, they are somehow getting the story wrong, missing some sneaky way in which their experimental system has outsmarted them. This concern sets the stage for a disagreement about whether the lab's results should instead be considered experimental artifacts. In examining the disagreement, I show that much hinges on how experimenters describe their results. Both problems thus provide windows into the multifaceted descriptive and interpretive work of making, doubting, or defending technoscientific claims.

"Moron Peer Review"

Of relatively recent origin, peer review is a constitutive practice for both granting agencies and academic journal publishing in the sciences and beyond. Yet the practice has been the subject of much debate over the past few years. A wave of high-profile retractions among a range of disciplines is partly to blame. Historian Aileen Fyfe argues that a little perspective serves to situate peer review between two erroneous perceptions: neither "shibboleth" nor "sacred cow," she contends, peer review is best "seen as a the currently dominant practice in a long and varied history of reviewing practices."[4]

For most of its history, journal publishing has depended heavily on the decisions of the mighty editor: first, single editors helming the communications of journals and, later, editorial collectives whose individual members judged content according to their specific expertise. In the eighteenth century, the dependence on the judgment of a single individual with limited expertise became a matter of concern for the learned societies to which journals were often attached. Leading French and British learned societies introduced new mechanisms for evaluating content in an attempt to remedy the problem. The Royal Society, which took control of *Philosophical Transactions* in the early nineteenth century, instituted a committee that decided on published content collectively by vote. The French Academie Royale des Sciences elaborated a different system, wherein paid academicians recruited committees to report back on discoveries and inventions by non-academicians in writing. These reports could then be used to persuade an editor to publish the findings.[5]

In the early nineteenth century, a philosopher of science named William Whewell suggested that the Royal Society commission reports from teams of prominent scholars on papers to be published in *Philosophical Transactions*.[6] The Royal Society was in the process of launching a monthly periodical, and these reports would be excellent fodder for the new venue. Whewell was borrowing heavily from the French model, yet his aim was not quality control, nor did he conceive of these reports as tools for making publishing decisions. Rather, he framed the reports as interesting documents in their own right that would unify the scientific enterprise in England while increasing its visibility.

Whewell's idea was enthusiastically accepted, and he signed himself up to write the first report in collaboration with the Royal Society's treasurer, an astronomer named John William Lubbock. The paper in question was by another astronomer, George Airy, and concerned some mathematical methods for calculating how the orbits of Earth and Venus were affected by the gravitational pull these two planets exerted on each other. Whewell and Lubbock both read the paper and instantly disagreed. They continued to disagree for months, Whewell having found the paper significant, Lubbock concluding that its equations were clumsy and inelegant. Whewell did not see a place for criticism in such a report and was concerned that a negative focus would repel other authors. Lubbock claimed he could not overlook the paper's errors.

Ultimately, having failed to agree, Lubbock took his criticisms directly to the author, including some suggestions for improvement. Airy, with some irritation, then wrote to Whewell that he would not change a thing. Lubbock swallowed his pride. Airy's paper was published, accompanied by Whewell's version of the report, with Lubbock a signatory. This was how the Royal Society launched into the practice of commissioning reports. Other British scientific societies soon followed suit.

In his retelling of this story, historian Alex Csiszar argues that Whewell and Lubbock represent two visions of what a report might do.[7] Whewell was the "authoritative generalist";[8] Lubbock the specialist, and a potential competitor. Whewell wanted to bolster the work and science in one breath; Lubbock wanted to list the work's demerits and bring them to the public.

Since Whewell initially won, at first the reports, all laudatory, were published. This lasted for little more than a year. The journal subsequently ceased publishing them, but it still commissioned them. They simply became anonymous and secret. Whewell's estimation of the reports' role changed as well. He now saw these reports as a means of defending the integrity of the Royal Society by excluding papers that were seen to be deficient, which meant that reports would have to include criticism. Csiszar chalks the secrecy up to two factors. First, the most readily available models for the emergent role of referee were the anonymous book reviewers who contributed prolifically to periodicals of the time.[9] The anonymity of these critics was thought to bolster their "oracular authority."[10] Second, and relatedly, there were the British norms of scientific civility: "Signing one's name to explicit criticism of a colleague would have been ungentlemanly."[11] Indeed, as Csiszar explains, "Virtual witnessing through the printed page had always depended on gentlemanly trust. By appointing judges whose job it was to take a suspicious eye to knowledge claims, the Society was throwing such gentlemanly conventions out the window."[12] Not surprisingly, then, within a decade of the emergence of the anonymous referee reports in England, suspicion about the motives of referees and incipient disdain for the new institution also began to foment.[13]

Nevertheless, by the middle of the century, the practice had become the norm among many learned societies, and by the century's end, the referee system had morphed into a "sort of universal gatekeeper with a duty to science."[14] The term "peer review" emerged only in the 1960s, accompanying a vast expansion in the practice's use for various disciplines, journals, and granting bodies.[15] Noting that the phrase "the scientific community" also dates to this period, Csiszar contends that peer review's renaming and expansion during this period reflects scientists' efforts to preserve research autonomy while maintaining their access to the gargantuan amounts of government support that had become available to them after World War II. The desired mixture of independence and dependence required a guarantee in the form of an internal mechanism for quality control. These were therefore the years when refereeing was made into a "symbol of objective judgment and consensus in science."[16]

Recent decades have seen a fair amount of debate about the reliability of peer review. The burden on referees, the psychology of bias, the tendency to block innovation and reward incremental steps, these are some of the grounds on which the centrality of peer review to so many scientific institutions has been questioned. One might assume that those whose success it has propelled would be more disposed to view it in a favorable light. Yet some very high-profile editors and scientists have written searing critiques that highlight some of the practice's perceived flaws. Richard Smith, former editor of the *British Medical Journal*, describes what he calls the "classic" peer-review system:

> The editor looks at the title of the paper and sends it to two friends whom the editor thinks know something about the subject. If both advise publica-

tion the editor sends it to the printers. If both advise against publication the editor rejects the paper. If the reviewers disagree the editor sends it to a third reviewer and does whatever he or she advises. This pastiche—which is not far from systems I have seen used—is little better than tossing a coin, because the level of agreement between reviewers on whether a paper should be published is little better than you'd expect by chance.[17]

The famed molecular biologist Sydney Brenner (mentioned in the introduction) voiced his own disdain for the practice in a piece for the journal *Current Biology* tellingly titled "Moron Peer Review."[18] In it, he detailed the process of applying for a grant and concluded by offering his own sarcastic guidelines for referees: "If it is novel and nobody knows whether it will work or not, call it 'over ambitious and superficial'; if it offers a better way of approaching a problem, protect all established plans by calling it 'unnecessary and redundant'; and if you find that the applicant has never done an experiment on the 8th base of tRNA, say he lacks the 'necessary experience to conduct these notoriously difficult experiments,' and turn it down."[19]

At one level, Brenner's gripes rehearse familiar themes. And one could just as easily cull the written record for defenses of peer review or dispassionate sociological analyses underwritten by data.[20] Yet the gimmick Brenner uses to deliver the message provides a helpful interpretive lens through which to view the practice, especially when it has led to a negative outcome. The profusion of euphemisms suggests that peer review can work in code, stoking interpretive possibilities and also a good dose of suspicion.

The Exchange

As artifacts of the publishing process, the three documents generated through Hecht et al.'s exchange with the journal (draft paper, referee reports, rebuttal) all provide insight into the anatomy of peer review as a textual practice. The primary document, the draft paper, manifests the distant professionalism of the scientific paper genre that largely divorces scientific findings from the foibles of human agency. As Rheinberger nicely sums up, "The author is grammatically silenced."[21] The reports and rebuttal, on the other hand, while still professional performances, are much more attuned to the subjective. Here and there, portions of the texts even teeter on the edge of incivility.

The reviews and rebuttal unfold as follows: The three reviewers all agree on one point, and that is that, if the synthetic genes are in fact rescuing the auxotrophic bacteria, this would be a significant finding and merit high-impact publication. They therefore address their doubts not to the significance of the project as a whole, but rather mostly to the extent to which Hecht et al. have sufficient evidence to substantiate their claim that their synthetic proteins are responsible for the rescue. Much of the doubt revolves around the question, "*How* is rescue achieved?" The first reviewer relegates this question to the end of a laundry list of concerns, but then frames it as the "elephant in the room." The second reviewer raises the issue by submitting an alternate theory: What if overexpressing the de novo proteins puts the cells under stress, which, the reviewer explains, "we know" activates stress responses that change the function of proteins in the cell? The reviewer continues that, most importantly, Hecht et al. should have purified the de novo proteins capable of rescue and tested their enzymatic activity, and calls the absence of such experiments surprising and disconcerting. The third reviewer concurs that a definitive link between the rescue and the activity of the de novo proteins seems premature, offering a numbered list of alternative accounts. The problem seems to be, then, that without fuller knowledge of how the de novo proteins function, Michael and his lab members might be under the mistaken impression that they have purposely produced and accounted for an effect that is in fact an accidental by-product of their experimental approach.

In the midst of this skepticism about how the rescues are achieved, a collateral set of concerns shoot out like pinballs. These collateral concerns easily fold into narrative passages or bullet points providing a false sense of continuity, yet constitute entirely different lines of attack. Reviewer two suggests extra caution because the area of de novo enzyme design has seen high-profile retraction in recent years; reviewer three finds the result pathbreaking, but notes that their own enthusiasm was dampened by the perceived pompousness of the style in which the paper is written, which the reviewer finds grating. The first editorial suggestion in a subsequent numbered list, therefore, is to "stick to the facts!" Not surprisingly then, the rebuttal must nimbly maneuver through different kinds of criticism. In doing so, it makes explicit that Hecht et al. read the reviews as challenging the lab on moral grounds and, moreover, that they see this challenge as having played a major part in the rejection of the paper.

Excavating Virtue

The rebuttal begins by pointing out that the authors, too, were astounded by the results at first, three years earlier, validating the incredulity of the referees. In the intervening years, therefore, lab members had done a lot of experiments to confirm the results. In the interest of keeping the paper short, an aesthetic marker of parsimony among many fields (anthropology not being one of them), mention of these experiments had been omitted from the paper. Although Hecht et al. sympathize with the reviewers for having been skeptical, given that these various validating experiments were omitted, their initial submission of the paper without mention of these further experiments suggests that they had not anticipated the reading their paper received. A generous reading would perhaps have taken obvious validating experiments and controls as a matter of basic competence and trust. In the absence of such trust, Hecht et al. assert competence by noting that multiple individuals conducted the experiments over several years: "We did every experiment we could think of to rule out artifacts. These experiments were performed by three different students, and were repeated over the course of three years. . . . We are confident that the results are real and reproducible."

The rebuttal then crescendos toward its most confrontational passage, in which it addresses the absence of trust head-on. The confrontation is framed as sympathetic to the frustration of reviewers, who, Hecht et al. fear, have read moral deficiency in and between the lines of the requisitely passive-voiced draft paper. The passive voice, in this sense, multiplies the possibility for readings that instantiate their own kind of hermeneutics of suspicion, that seek the hidden subjects behind the carefully orchestrated subjectlessness. The potential moral deficiency, in this case, revolves around the reasons for the absence of experiments with purified proteins, which would have allowed Hecht et al. to figure out how exactly the proteins rescue the sickly cells. Hecht et al. respond:

It would be *foolish* or *arrogant* to present this work without attempting to purify the proteins and assess their activities in vitro. In fact, we spent nearly three years working with the purified proteins (and cell lysates). However, this work is considerably more difficult than we (or the referees) expected, and it is not yet complete. Therefore, we omitted these studies from the submitted manuscript, and planned to continue this work for eventual publication in a later paper. Unfortunately, by failing to mention this work in the submitted manuscript, we may have given the impression that we were either too *lazy* or too *careless* to attempt such experiments. Nothing could be further from the truth. We are eagerly pursuing the biochemical experiments, and will certainly address this in the revised manuscript.[22]

The passage above is striking for a simple reason. The defense mounted in it makes little difference when it comes to experimental results; the omitted experiments are unexpectedly difficult and hence ongoing. The passage is therefore not about shoring up experimental results but about clarifying the moral significance of their omission. And it suggests that these moral considerations are of paramount importance, having perhaps torpedoed the paper's publication. The remediation of a reading that suggests moral deficiency requires the recalibration of the collective moral persona, performed in text, and this persona, in turn, plays an overarching function in the paper and beyond, guaranteeing the trustworthiness of the lab and hence the entire network of experiments and claims. Of particular concern are a set of potential collective vices, excesses, and deficiencies—arrogance, foolishness, laziness, carelessness—qualities that might be as damning for the paper's publication in Hecht et al.'s assessment as the actual absence of the results of protein purification experiments. Notably, the expression of eagerness is therefore a key antidote to perceived vices, like laziness and carelessness.

In his 1994 book *A Social History of Truth*, Shapin argues that the history of truth is a social history because "what we know about the world is arrived at, sustained, and recognized through collective action."[23] Against the romantic view of the lone knower, Shapin argues that knowledge is inherently a group endeavor predicated on moral bonds that both enable social life and also provide the vital infrastructure for the circulation of information that cannot be obtained firsthand. Shapin proposes the term *trust* as the one most suited to the task of describing this bond. It is trust that knits the social and epistemic orders together.

Trust, notes Shapin, has a morally consequential, and a morally inconsequential, variant. Expectations about how the world works, arrived at through induction, as when we trust that kicking a ball will send the ball flying into the air, are morally inconsequential. The morally consequential variant of trust, which is the one with which Shapin is most concerned and which bears on my analysis here, refers to the kind of trust the disappointment of which activates moral approbations such as blame.[24] Such trust also has important epistemological dimensions. Not only do our relations with those around us depend on the predictability of their actions, but also we rely on trust insofar as so much of what we know is in fact a reflection of what we have been told, and therefore implicates our relationships with others.

Shapin notes that the relationship between trust and the social order has been the topic of much philosophical and scholarly interest.[25] In contrast, he observes, the relationship between trust and knowledge production had been notably de-emphasized.[26] This is not a problem of simple neglect: "Rather," Shapin explains, "much modern epistemology has systematically argued that legitimate knowledge is defined precisely by its rejection of trust. If we are heard to say that we know something on the basis of trust, we are understood to say that we do not possess genuine knowledge of it at all."[27] Distrust, in the guise of scientific skepticism, is in fact foundational to a certain story about legitimate knowledge, despite the fact that distrust can pose a threat to the social order. Yet, even in the sciences, direct experience of all the necessary components that underlie a claim is simply impossible, and trust is used to fill in the gaps. In fact, Shapin argues, "distrust is something which takes place at the *margins* of trusting systems."[28] Given that textual media are one of the ways in which the imagined communities of technoscience are constituted, we might then ask how trust figures in and through text? The question becomes more compelling if we keep in mind the impersonality of so much scientific writing.

To pursue these questions, we return to the rebuttal. The eagerness recruited as the antidote to the felt charge of laziness is not to be confused with "enthusiasm," which is the subject of a separate section in the document. In it, Hecht et al. reply to the criticism of their paper's grandiose prose by conceding that their own "over-abundant" enthusiasm diminished the enthusiasm of the referees. They acknowledge that this might be an irritant for readers more generally and promise to "curb our enthusiasm" in a revised draft. Shapin, again, proves helpful, this time through his contributions, with Simon Schaffer, on the topic of modesty.

In Shapin and Schaffer's classic book *Leviathan and the Air-Pump*, the authors famously argue that Robert Boyle's experimental philosophy, a precursor to the modern experimental tradition, determined facts by aggregating the individual beliefs of witnesses.[29] The experiment was thus a kind of trial, subject to the fallibility of witnesses whose judgments nonetheless gained strength in numbers. Not only the knowledgeability but also the moral constitution of the witness who provided testimony mattered.[30] The scientific text, therefore, was encumbered with making the virtue of witnesses legible, thereby convincing the reader of the admissibility of testimony.[31] What were the virtues in question? Shapin and Schaffer argue that modesty was a central quality of Boyle's scientific writings: "A man whose narratives could be credited as mirrors of reality was a *modest man*; his reports ought to make that modesty visible."[32]

Modest presence, in turn, was orchestrated through a number of textual techniques. First, the experimental essay form lent itself to modesty, and was contrasted with the philosophical system. As Shapin and Schaffer explain, "Those who wrote entire systems were identified as 'confident' individuals, whose ambitions extended beyond what was proper or possible. By contrast, those who wrote experimental essays were 'sober and modest men,' 'diligent and judicious' philosophers, who do not assert more than they can prove."[33] Modesty was also conveyed through style. Boyle eschewed "florid" style for a "naked way of writing."[34] "Moreover," continue Shapin and Schaffer, "the 'florid' style to be avoided was a hindrance to the clear provision of virtual witness: it was, Boyle said, like painting 'the eye-glasses of a telescope.'"[35]

In her own repurposing of Shapin and Schaffer's "modest witness," Donna Haraway has observed a strange quality at the heart of these stylistic signs of modesty.[36] She writes, "In order for the modesty . . . to be visible, the man— the witness whose accounts mirror reality—must be invisible, that is, an inhabitant of the potent 'unmarked category,' which is constructed by the extraordinary conventions of self-invisibility."[37] For Haraway, self-invisibility, against which she rightfully fumes, is central to the founding of muscular, masculinist, modern science. Yet, despite its centrality to the authority of the modern sciences, self-invisibility might not be a pre-perfected subject position but rather a target that can be missed, or rather a performance replete with the kinds of difficult contortions the term "self-invisibility" would seem to suggest and that is subjected to disciplining practices. The exchange with the journal demonstrates an almost parodic hermeneutic exercise in which authors and readers have to depth-mine for the subjective, which is mostly hidden from view in the draft paper, in case what has gone wrong is not the science but the purposely obfuscated ethical guarantees of the testimony needed to substantiate it. The obfuscation, further, is policed: "Stick to the facts!" That is, reviewers read for the moral constitution of authors. By convention, that constitution is hidden from view; its being hidden is a key part of that very same moral constitution. The subjective is then brought to the surface, yet in places where it is already on the surface—a demerit—it is pushed back down to the depths. In this way, peer review subjects authors to fairly intricate textual-cum-moral discipline.

Two Stories about Trust

From antiquity through about the second half of the seventeenth century, assessing the truth of testimony, and avoiding the twin evils of gullibility and skepticism, constituted a major philosophical problem.[38] Ways of handling knowledge gained through testimony and authority occupied a great many luminaries through the ages, including Aristotle, Cicero, Augustine, and Locke. From the late seventeenth century, we can trace two paths to the present that suggest the subsequent fate of trust and authority. The first path follows the metamorphosis of trust from something that transpires between individuals to something that is impersonally endowed in institutions. Premodern societies, the story goes, relied on face-to-face interactions to assess credibility. In such interactions, familiarity comes to play a constitutive role in forming judgments about the veracity of knowledge gained through others. As Shapin puts it, "Premodern society looked truth in the face."[39]

Times have changed. Today, "the village has given way to the anonymous city," and trust is bestowed on institutions and abstract entities.[40] That is, trust has morphed, but it persists. And it has not contracted, but rather expanded its social and epistemological role. This is because modern life is endlessly reliant on a kind of trust Niklas Luhmann calls "system trust," which reduces complexity. As Luhmann writes, drawing an example from the economic sphere, "anyone who trusts the stability of the value of money, and the continuity of a multiplicity of opportunities for spending it, basically assumes that a system is functioning and places his trust in that function, not in people."[41] Shapin summarizes, "In other words, modernity guarantees knowledge not by reference to virtue but to *expertise*. When we give our trust to—'have faith in'—modern systems of technology and knowledge, our faith is now widely said not to be in the moral character of the individuals concerned but in the genuine expertise attributed to the institutions."[42] Institutions, in turn, constrain individuals in such a way as to earn public trust. In the middle of the twentieth century, Robert Merton expounded the view that institutions produce the policing practices and incentive schemes that constrain individual freedom and guarantee objectivity. Shapin sums up, "The gentleman has been replaced by the scientific expert, personal virtue by the possession of specialized knowledge, a calling by a job, a nexus of face-to-face interactions by faceless institutions, individual free action by institutional surveillance."[43]

Granting the power of this narrative in explaining something of the relation between publics and experts, Shapin qualifies it when it comes to what goes on *within* expert communities. Here, he asserts that the world of familiarity and virtue, of the face-to-face variety, is far from lost, an assertion he substantiates in his more recent book, *The Scientific Life*.[44] By circumscribing epistemic communities as groups for which personal trust still matters, Shapin sidesteps the "system trust" story. How "inside" and "outside" relate to each other therefore remains something of an open question. As Paul Rabinow has argued, the demarcation of "inside" and "outside," and the stabilization of forms that maintains these demarcations, has been a major accomplishment of the sciences. That is, the sciences have been successful in "developing forms that compartmentalize and maintain strongly demarcated boundaries between the domains marked as external world (on which [they are] dependent for money, political support, training facilities and many other things) and [their] own well-policed and well-demarcated internal domains."[45] In this light, peer review can perhaps be understood as one among a set of practices—neither the oldest nor the newest—that facilitates the production and maintenance of "inside" and "outside." Or, in Shapin's language, peer review is situated at the intersection of the personal and the institutional, of the "gentleman" and the "expert," where one transubstantiates into the other.

Yet it seems worth noting that trust in peer review hardly translates into public trust in synthetic biology. The relationship between synthetic biology and "the public," of which much has been made by practitioners and observers alike, finds more pressing points of contact, at least rhetorically, in the need for "public engagement." This is because concern with public trust in this domain seems to depend less on the mechanisms for guaranteeing truth or maintaining research quality and more on problems determining who should decide how new technologies will be used, whether practitioners should be pursuing such lines of research in the first place, and whether existing institutional forms and venues can be made to accommodate such questions.[46] The concern with "public engagement" shows that "inside" and "outside" can be related to each other in different ways, which, in turn, rely upon, or call into existence, different forms and practices for maintaining or crossing these boundaries.

In his book *A History of Reasonableness*, Rick Kennedy traces a different path from the seventeenth century to the present than the one involving the development of "system trust."[47] Some of Kennedy's concerns echo Shapin's: Kennedy argues that testimony and authority continue to enjoy an important, if underappreciated, role in knowledge. Importantly, however, in Kennedy's view, it is the skill set used to judge them, which he terms "reasonableness," that has fallen into disrepair. In other words, trust continues to matter, but without a proper epistemology.

Kennedy blames Kant. For Kennedy, Kant heralds the beginning of the end of a robust and enduring educational tradition, replacing an epistemology of testimony with the celebration and cultivation of self-reliance in the guise of what would eventually become the "critical thinking movement," which furnishes Kennedy with his ultimate target.[48] Kennedy's narrative and periodization raises the possibility that the collective witnessing that guaranteed experimental knowledge might have triumphed just as the skill set from which it drew its strength was about to weaken or disappear. The question of whether trust has ceased to exist might then be replaced by the observation of the richness or poverty of the tools available for handling it.

The reader may still wonder, given that Michael is an old hand at paper writing and publishing, wouldn't he know the requisite performance? In other words, why so visible? In this case, the highfalutin rhetoric, which is read as "over-blown," may be more than a misstep in the crafting of a moral persona. It may have something to do with the lab's betwixt and betweenness.

The style presented by key figures building up synthetic biology has tended toward the highfalutin. Venter and the more mediagenic members of the engineering community all dabble in grandiose prognostications and pronouncements about synthetic life. Teetering on the boundary between discovery and commercialization, these scientist- or engineer-entrepreneurs rely on charismatic authority to deliver a sense of urgency or to begin to package and sell a product. Sometimes, they are allowed a little wiggle room in formal, high-impact scientific papers. More often, they are handed megaphones in collateral spaces like review articles, keynote speeches, and interviews.

Their claims to grandeur thus erupt from the sidelines, the access to which is predicated on some celebrity in the first place. The grandiose rhetoric, however, cannot necessarily be engaged in by others or returned in kind.

What is more, Hecht et al. enter into this rhetorical domain in order to explicitly claim their own contribution to a truly fabricated biology as more radical than some of the more dominant and highly visible approaches like the parts-based approach. Yet notably, none of the referees even mentions synthetic biology. One of the referees suggests the paper is too narrow for a broad readership and should be sent to *Protein Engineering, Design, and Selection*, a specialized journal for protein scientists, effectively reading the paper as a contribution for specialists. A group of engineering-oriented synthetic biologists, or chemical engineers-cum-synthetic biologists, would likely have written very different reviews, confirming that "peer" is a very broad, and malleable, category. Richard Smith, the former medical journal editor mentioned earlier, neatly summarized this point by asking, "who is a peer? Somebody doing exactly the same kind of research (in which case he or she is likely a competitor)? Somebody in the same discipline? Somebody who is an expert on methodology?"[49] How the editor answers the question will obviously change the content of the reviews.

It just so happens that chemists present a particular challenge when assessing the research findings of Hecht et al. because of their fairly consistent (in Michael's accounts) insistence on knowing *how* the proteins work. Michael framed this interest in function as a collective cultural trait. He once explained it to me this way: for chemists, the main question is "What do these proteins do?" rather than "What can you do with these proteins?" Whenever Michael gave a talk to an audience of chemists, he noted with some frustration, the question of function stole the show.

Experimental Artifacts

This brings us to the crux of the reviewers' concerns. Absent an account of function, all three reviewers propose alternate causal pathways that would implicitly render the lab's results *experimental artifacts*—phenomena generated or affected by the experimental setup in unintended ways. Yet only one reviewer raises this possibility explicitly, setting the stage for a disagreement about what constitutes an experimental artifact in the first place. The disagreement, in turn, sheds some more light on the dynamics of description and interpretation that accompany attempts to make, and stabilize, techno-scientific claims.

Though commonly invoked, experimental artifacts lack formal definitions in many fields. Informal definitions, however, are easier to come by, particularly for an anthropologist. I once posed the question, "What is an experimental artifact?" to a philosophy of science reading group—composed of scientists and philosophers—in which I take part. A physicist in the group quickly volunteered the following definition: They are "the things you don't mean to do that mess up your results." If material artifacts have been, since Aristotle, understood to be things made on purpose, experimental artifacts are colloquially understood to be things done by accident. They are things to be ruled out, controlled for, and carefully disposed of, to make room for believable results. Artifacts are often sources of local consternation, though occasionally they become problems for general reflection and even the subject of pronounced disagreement.[50] In the exchange, they prove fodder for a significant dispute, which reveals that there is more to them than is immediately apparent.

The disagreement over what sorts of things should properly be termed artifacts surfaces in the exchange after one of the reviewers raises a particular concern: What if the de novo proteins, rather than compensating for the deleted genes in the auxotrophs by providing the same activity as the deleted natural enzyme, are rescuing the bacteria by modifying the metabolic regulation of the auxotrophs in some unknown way? The de novo proteins, for example, might be binding other proteins in the bacteria, thereby changing their function, rather than compensating for the deleted function on their own. In this reviewer's estimation, such a circuitous path would qualify the results as artifactual.

In their rebuttal, Hecht et al. acknowledge the possibility that the rescue works via an alternate mechanism and note that the paper is agnostic on function. They in fact suggest that some of these alternate mechanisms might be among the more interesting possibilities. Yet Hecht et al. insist that such mechanisms would constitute a result, not an artifact. They write,

> We believe there is a misunderstanding about the meaning of "artifact." The designed proteins may have the same activity as the deleted natural sequence, or they may function by some novel mechanism, such as binding to a natural

protein and altering its activity. Either way, a protein designed *de novo* provides a biological function enabling cell growth where growth would otherwise not occur. This is not an "artifact." Artifacts pertain to contaminations, leaky phenotypes, etc. Biological function encoded by sequences designed *de novo*—regardless of the exact mechanism—is not an artifact.

The question articulated through the disagreement is not whether the results are actually generated through some unknown interference. The mechanism of rescue is not what Hecht et al. contest. What is at stake, instead, is the boundary marker between an artifact and a result, a boundary that turns out to be interpretively rich.

In his book on the philosophy of experimental biology, Marcel Weber provides one of the few general discussions of artifacts to be found in the philosophy of science.[51] In his discussion, Weber provides the following tentative definition of an artifact: "In laboratory slang, an 'artifact' is a phenomenon or appearance that is thought not to represent a real biological structure or process, but one created by the experimental method or instrument used."[52] Weber notes that this definition seems simple but is in fact problematic on further examination because *all* experimental results are in a sense created by the experimental procedure. This observation most certainly holds true for experimental artifacts in areas of research whose aims are predominantly technological, and in which human agency is the wanted outcome. In such areas, experimental artifacts pertain to questions about agency and control.

Here, words matter. The rebuttal renders the lab's results as follows: "A protein designed de novo provides a biological function enabling cell growth where growth would otherwise not occur." But the draft paper also uses a different term to describe the process. Where the rebuttal carefully chooses the term "enable," the paper sometimes uses the term "rescue." Suppose someone is trapped in a burning building. I smell the smoke, but I do not have a means of getting to the fire. I therefore pick up the phone and call someone who is closer to it, who braves the fire and rescues the victim. It seems like a stretch to credit me with the "rescue," though I undoubtedly "enabled" it. If I had generated the rescue plan and coordinated the effort my claim would be stronger. The reader gets the drift; we are in interpretive waters. What is at stake in the gap between "enable" and "rescue" is a set of questions concerning the relationship between causal chains and agency, and so "rescue" leaves room for some artifacts that "enable" does not. We might even say that the reviewer's main concern turns out to be an *artifact of description*, resulting from some fairly innocuous seeming differences in word choice.

Risks Worth Taking

In his work on experimental vitalism, Robert Mitchell distinguishes among three dominant science studies approaches to the study of experiments: an epistemological approach, which emerged in the early twentieth century and "focuses on the contribution of experiments to true accounts of the world"; a sociological approach, with its roots in the 1970s, which explores the ways in which experiments both "facilitate and resolve social questions"; and what he terms an *ontogenetic* approach, which has been more or less dominant in recent years and "focuses on the ways in which experiments bring *new* entities and assemblages into being."[53]

The first and the last of these approaches can be arrayed around different vectors of "experimentation." On the one hand, experimentation is an activity associated with rigidly held methodological tenets aimed at answering questions in rigorous ways. Accordingly, an epistemology of experimentation engages in an activity akin to calking a tub for leaks, "heavily weighted toward the wary avoidance of errors through experimental tests and evidentiary argument."[54] On the other hand, experimentation implies open-endedness, tentativeness, and flexible exploration. It suggests the innovatory, the risky, and the unconventional. As native a notion to science as it is to art in contemporary usage, experimentation suggests "things said and done just in case they may elicit fresh and even constructive responses, or flush old stagnancies."[55] An interest in this valence of experimentation foregrounds "the eager pursuit of the new."[56] It is much less interested in shoring up leaky tubs. The two poles pull in different attitudinal directions as well, resonating with tensions between stinginess and generosity, certainty and wonder.

My point, in delineating these two vectors, is not to make a case for the need to pay more attention to the latter aspect and aim of the experimental (techno)sciences, as others have successfully done, but to simply note that these aspects of experimentation can be viewed as resources for practitioners themselves. That is, many experimenters negotiate these poles in their own practice and also appeal to these contrasting sensibilities in others. From this angle, the rebuttal can be understood as an effort not to plug all potential leaks, but rather to remediate suspicion and thereby call forth a particular kind of social fact in the realm of peer review, the "generous reading," which is tolerant of a degree of uncertainty because such uncertainty is sometimes deemed worth the risk.

Chapter 5

ON THE MOVE

An ethnographic perspective on the activity of a lab calls special attention to the physical habitat within which lab members operate. Writing about scientists and the buildings in which they conduct their research, Peter Galison has noted that architecture serves as a "daily reminder to practitioners of who they are and where they stand."[1] Thomas Gieryn extends the range of actors and institutions for whom the built environments of research can be meaningful. He explains, "Through their very existence, outward appearances, and internal arrangements of space, research buildings give meanings to science, scientists, disciplines, and universities—for those who work inside and for those who just pass by."[2] In this final chapter, we therefore begin by wandering together through a few buildings on the Princeton campus.

Our wander is organized around an event. If the exchange with the journal analyzed in chapter 4 represents the attempt by Hecht et al. to find a venue for their research, this chapter details a more literal crisis of place. In March 2011, Michael's lab transitioned from the old chemistry building to the new one on the Princeton campus, a built ode to transparency. For lab members, the move from building to building catalyzed a set of experimental failures. Cells that were expected to grow did not. Petri dishes that were supposed to be covered with lively spots emerged from their growth conditions after a few days with nary a speck of life. In the new building, experiments stopped working for unknown reasons. Thus, the move, and the experimental perturbations it produced, furnishes us with one final orthogonal entry point.

Experimental methods are, of course, built around institutionalized repetition aimed at proving reproducibility. But beyond explicit attempts to validate results, there are the myriad forms of repetition that are spurred by circumstances, and that are not, strictly speaking, aimed at confirmation. Such pragmatically motivated, opportunistic repetition in experimental practice is nevertheless integrated into the epistemic infrastructure of laboratory life, through which practitioners learn over time—through hints and intuitions, surprises and disappointments—whether to trust their own results. The move to the new building is a case in point, serving not only as "a reminder to practitioners of who they are and where they stand" but also as a $298 million incidental experimental control.

From Frick to Frick

The original Frick Laboratory, home of the chemistry department for many years, is a Gothic-style building, with arches and vaults, named for Henry Clay Frick, turn-of-the-century industrialist, chairman of the Carnegie Steel Company, and noted arts patron. Frick's initial interest lay in funding a law school for Princeton in the 1910s, but his support was redirected to the construction of a chemical laboratory. Plans for construction hit a snag when Henry Frick saw the price tag of the project, supposedly, and ironically, inflated by the price of steel. Frick died in 1919, leaving the university a sizeable gift for discretionary use. Other needs having been found more pressing, Frick's donation was not ultimately used in the construction of the laboratory, which was financed by the Princeton University Fund and built in 1929. Recalling his interest in the project, the trustees voted to name the new building after the great American industrialist.[3]

Behind Frick Laboratory stands Hoyt, the very functional and relatively unremarkable 1970s extension of Frick, built of machine-made concrete and glass. Hoyt contains many labs and is connected to Frick by walkways and hallways. Michael's office and lab were located in Hoyt, while lab meetings took place mostly in Frick. Frick was slated for massive renovations that would transform it into the home of the economics department. The name Frick was to migrate down the hill to the new site of the chemistry building, leaving room for a new donor's name on the face-lifted Old Frick.

Construction of New Frick was completed in late 2010. Its façade hangs back from Washington Road, down the hill from Old Frick, in a scenic and shady area of the campus. On a cloudy day, New Frick looks slightly dour, aged before its time. Covered in dark, grate-like panels that filter light into the building, the exterior manifests some of the hallmark industrial banality of 1960s university buildings, with a key difference. The 216 panels covering the façade are in fact photovoltaic panels, power generators that simultaneously shade the glass-roof-covered interior atrium.[4]

And what an atrium. Four tall stories high, twenty-seven feet wide, this large hollow interior is spanned by pedestrian bridges on all levels, while aerial bubble sculptures riffing on molecular themes hang from the glass ceiling. At the dedication event for the building, an architect at the firm responsible for the building's design explained that this large central space was intended as a sort of street. On both sides, the atrium, which spans the length of the building, is bordered by glass, beyond which one sees workspaces: labs on one side, group meeting rooms and classrooms on the other. The new building is a monument to transparency.[5]

In his essay "Architecture of Transparency," Emanuel Alloa provides a history of architectural transparency that resonates with some of the aesthetic effects of New Frick.[6] His analysis begins with the twelfth-century Gothic style and its instantiation of "vertical transparency," a topic that would be admittedly far afield from my own, were it not for the collegiate Gothic style that reigns supreme on the Princeton campus. Enter Princeton's Firestone Library, with its imposing Gothic façade, located at the northern edge of the Princeton campus, and you will find along the main stairwell a piece of Oxford University. Embedded in the wall of the wide, marble, B staircase is a stone slab, "From Pembroke College Oxford founded 1624 The College of Doctor Johnson." The architectural style that characterizes most of the campus is the Oxbridge-inspired Gothic Revival, selected as the style of choice just after 1896, when the trustees decided that what was then the College of New Jersey should be a top-tier research institution. A nineteenth century characterized by unplanned and heterogeneous construction on the campus was therefore followed by a concerted effort to mimic Oxford and Cambridge through Collegiate Gothic.[7]

The choice of style was symbolic of more than just academic excellence. The historian Johanna Seasonwein links the revival of Gothic style at Princeton to broader religious and class anxieties in America at the turn of the previous century. In particular, she observes, "Members of the old-stock Protestant Bourgeois and upper classes embraced medievalism and other 'antimodern' sentiments as a way to combat their growing unease with the changes wrought by industrialism, immigration, and later, war. The later Middle Ages were seen as a purer, more child-like, and more moral time that could provide a model for reviving the ethic of work and self-control that Protestants embraced."[8] Since the threat to this Protestant order was coming largely from the influx of Catholic immigrants, the use of Gothic symbols amounted to a co-optation of a Catholic architectural vocabulary toward American Protestant ends.

In his essay, Alloa explains that the Gothic period saw the emergence of architectural techniques involving the use of pointed arches that allowed for a merged, light-filled central space. This space enabled communion with the unhindered light of revelation, exemplified in the Gothic Basilica of Saint-Denis, classically interpreted by Erwin Panofsky as symptomatic of the "principle of transparency."[9] The "street," with its glass-filtered light and massive size, activates some of these divine associations. Ron McCoy, Princeton University architect, made the associations explicit when providing an assessment of the atrium: "That is probably the singular most stunning interior space on campus since the chapel—in terms of its size, the kind of grandeur of it. People are going to drop their jaws and look up when they enter that atrium, just as they do in the chapel."[10]

If awe and the ascending gaze suggest sacred space, with a touch of antimodernism, then transparency, in New Frick, is formally couched as a way of engendering the more secular ends of openness and collaboration. Again, McCoy explains, "The scientists are now going to be working in this remarkable glass loft, and they're going to have views through the building, but always with a sense of openness. The way that the sense of openness captures the spirit of collaboration is hugely important."[11]

The "spirit of collaboration" is built into New Frick in more ways than one. The new building is the direct result of close ties between Big Pharma and university-based basic research.[12] The story goes like this: In the 1990s, Princeton chemist Edward Taylor approached global pharmaceutical giant Eli Lilly with a possible cancer drug. This was the beginning of a relationship that was framed as exemplary of the potential benefits of collaboration between basic research and Big Pharma in Princeton's wide circle of student, institutional, and alumni publications. The patent for the resulting drug, Alimta, which is exclusively manufactured and marketed by Lilly, belongs to Princeton. A Princeton University research news publication launched its article on the collaboration in 2011 with the following praise: "The story of Alimta® highlights how university research can yield benefits to human health and society while enhancing revenue at a major pharmaceutical company."[13] The phrasing has the feeling of a two-birds-with-one-stone claim, which obfuscates some of the strangeness of that second bird—enhancing revenue at a major pharmaceutical company is not an obvious achievement to tally for a university. Then there is the hidden third bird: namely, the fact that Alimta proved quite the revenue stream for the university as well. The transparent building was financed entirely through Alimta royalties.

The Lilly license is irrevocable and exclusive for the lives of the patents.[14] The patents were, in fact, set to expire in 2011, but were extended by the US Patents and Trademark Office to 2015, as a result of delayed approval by the FDA. Princeton and Eli Lilly have teamed up as co-plaintiffs on at least three occasions to impede the production of generic and variable versions of the phenomenally expensive drug. The extension and generics litigation angered critics, who questioned Princeton's incentives. According to Lilly's financial statements, the company sold $1.15 billion of Alimta in 2008 alone. Princeton is contractually entitled to royalty payments equaling a "single-digit" percentage of net sales. Princeton, with true preppy propriety, had declined to discuss its paycheck from the drug, but the new building's price tag was estimated at a whopping $298 million, a figure still far exceeded by the estimated cost of research and development for the drug. Of our new problems, this one is an old one. The intricacies of relations between corporations generally—Big Pharma in particular—and universities are well beyond the scope of this book. But, we may still ask, what happens when an institution of higher learning—a nonprofit explicitly committed to "the service of humanity"[15]—is a litigant and on the side of Big Pharma? One result is some awkward architectural transparency.[16] In this case, religious-corporate-academic mixed-messaging involving transparency, divine light, openness, ownership, knowledge, technology, and profit, applied to the new home of the chemistry department, instantiates an ideological mash-up in built form.

Not surprisingly, transparency lends itself to surveillance, as "visibility as a conduit for knowledge is elided with visibility as an instrument for control."[17] As I wandered around the third floor one afternoon, I focused my attention on the grad student offices across the atrium, one floor below me. I zoomed in: One graduate student at his cubicle, his back and shoulders visible, ergonomic Danish swivel chair (this is Princeton, after all) turned slightly away from the window. His computer screen was directly in my line of vision. If he turned his laptop to obscure my view, his screen would be visible to his associates in the room. This would be perhaps slightly creepy, but on the whole irrelevant, were it not for a particular history of student surveillance in the sciences, complete with its own set of stereotypes and mythologies. Among these stereotypes and mythologies, the field of chemistry is particularly notorious. Michael once explained to me that organic chemists have the reputation of being the most brutal when it comes to demanding long hours of their students and making sure that they get them. He recalled a colleague once telling him that his PhD adviser and lab head had expected sixteen-hour workdays from his graduate students. The colleague in question would show up in the morning, drop off his backpack, and go get breakfast. One day, the adviser pulled him aside and said, "Your backpack doesn't run experiments."[18]

Hugh Gusterson has recently written a rousing and timely call for anthropological studies focused on the university, lamenting the fact that, while other fields have taken up the university, the topic lies fallow among anthropologists.[19] Gusterson suggests some possible causes for this omission—a sense of professional discretion about anthropologists' own institutional abodes among them—yet his explanation of why universities should figure more prominently in anthropological work renders their richness mostly in terms of critiques of neoliberalism, capitalism, and empire. These are certainly important frameworks through which the relevance and potential nefariousness of the university can and should be apprehended. Yet when placed in the particular purview of these concerns, the kinds of problems and situations in which the university could come to play a role become perhaps a bit too limited. University life—hierarchical, "meritocratic," bureaucratic, micro-political—builds careers and ruins them, situates, sorts, prioritizes, and normalizes, rewards and punishes, sometimes in subtle ways, though equally frequently in ways so blunt as to elude attention. Moreover, these dynamics seem worth attending to, both because they invite the reflexivity anthropologists lament not having internalized in relation to their own organizational and institutional lives, and also because they play a big role in shaping knowledge and technological regimes.

In the case of Hecht lab, the excitement over new fancy facilities was significantly tempered by institutional politics that also figure strangely in the new transparent milieu. The ribbon-cutting ceremony and inaugural festivities took place in the late spring, as students were wrapping up their senior theses. For the duration of the celebration, undergraduate students were not allowed in the building, barring them not only from the booze and hors d'oeuvres, but also from their research. The reaction to this exclusion from Hecht lab members was a collective groan that conveyed the sense that the gesture was tactless.

Though it was a minor episode in terms of its effect on laboratory life, the brief expulsion of the undergrads added to the collective deflation arising from a bigger one. The move had given implicit department hierarchies new spatial expression, making them all too visible. Michael's lab was actually losing space in the move, even though New Frick was designed to accommodate many more benches than Old Frick and Hoyt combined. Much of the better real estate (closer to the outside-world-facing windows), I was told, was intended to attract future faculty and so, for the time being, was being kept unoccupied.

Moreover, in the old building, the laboratories were discrete rooms. The spatial boundaries of the lab had therefore coincided with the walls. New Frick was built around an open lab concept, in use since the 1960s and in vogue since the 1980s or so.[20] With open floor plans and lots of empty lab space, the boundaries in New Frick were largely immaterial, which gave more emphasis to the delimiting exercise from institutional higher-ups, on the one hand, and to the requirement that lab members regulate those limits themselves, on the other. The intended openness therefore turned the lab decidedly inward. Since their assigned lab space on the third floor was not yet ready for occupancy, Michael's lab was put in a temporary workspace, technoscientist squatters, defiantly spreading their belongings and instruments beyond the allotted ten benches.

On my way to my first lab meeting in New Frick, I was ascending the zigzagging staircase to the third-floor meeting room with Betsy when I recognized some more members of Michael's lab across the atrium through the glass. I waved at Maria, a graduate student, and was surprised to find out how quickly she noticed and waved back. Visibility, in New Frick, is a two-way street.

"Floors" and "Ceilings"

Moving, Michael tells me, is a drag. He's done it twice this year: once from home to Forbes College, an undergraduate dorm where he serves as "head of college," and once from Frick to Frick. The relocation to New Frick required a fair amount of preparation from lab members. Packing, labeling, more labeling, finishing experiments, and cleaning up workspaces; then unpacking in a new lab with different dimensions, different layout; setting up, getting oriented, mixing new compounds and culturing new cells. Weeks of potential research time go by in this way. All of this work had now been completed, and the first few rounds of experiments had finally been attempted in the new surroundings.

At the lab meeting, Sarah presents her most recent round of work, which was soon to culminate in her senior thesis. Sarah, I discover, like many of her colleagues, has been unable to replicate results in the new building. In New Frick, experiments had just stopped working—experiments whose results had already been published.

In an interview, Betsy provided a more detailed account of the month's eventful experimental nonevents:

> We have these five strains where we have synthetic proteins that rescue them. CisD was a real auxotroph up there [meaning in Hoyt], and [Sarah] got these rescues, and it was all good, and it was a fifth one, yay! And so now we have these five strains. Very replicable that they rescue on minimal media. We come down the hill, Sarah is working with CisD, Charlotte's working with SerB, Maria is working with GlitA, and I'm sort of doing stuff with all of them, and suddenly we start noticing that we're not getting rescue where we should. . . . And so we're like, this is unfortunate. I mean because most things are still the same. We still have the same bottle of salt; we still have the same glassware . . . So Sarah and Charlotte and I talked about this, and then I decided to take all five strains and just test them . . . with the caveat of, I did not include the positive positives, which is the natu-

ral protein. And certainly non-auxotrophic cells will still grow on minimal media. And also things are growing totally fine on rich media. . . . It was like a Friday or something, so I plated all these things, and by Wednesday group meeting none of them had grown. And they should have grown in five days. So we bring it up at group meeting. We had a whole conversation and raised the possibility of killer phage or something, but that seemed unlikely because our things were growing in rich media, and it should just kill everything. And so, I don't know, I was literally having a panic attack, because this is kind of what I'm working on right now, so if it doesn't work, at least my name is not on the paper [laughs]. I'm pretty sure Michael was having more of a panic attack than me.

Sarah's presentation at the lab meeting is deflated in tone and uncharacteristically short. She has been running negative controls on CisN, one of the newest confirmed rescues that she had discovered. But now that other experiments aren't working, she doesn't know how to interpret these controls either.

In one of his more didactic lab meeting moments, weeks earlier, Michael had explained to the group that, in order to do these experiments, you need to have a "floor" and a "ceiling" (he draws two dashes on the board, one hip-height, one head-height). Floors and ceilings are experimental bearings. They provide the basic criteria by which meaningful effects can be judged and later defended. The floor is usually provided by the negative control. It is the absence of growth seen when the sickly bacteria have neither their missing gene nor the lab's synthetic protein to depend on. The ceiling is the positive control or the growth that would be expected under "normal" circumstances. Years into the auxotroph research, these were still not fixed coordinates, since parts of the experimental system were not yet fleshed out. For example, "ceilings" are defined as normal growth, yet what constitutes normal growth for abnormal cells? Is it the growth one would expect from healthy *E. coli* or the growth expected from auxotrophs that have had the wild-type protein reinserted? The latter experiments were on the agenda, but had not yet been done. "Floors" and "ceilings" are built through and alongside experimental systems.

In Sarah's case, the negative controls were not growing, which, under normal circumstances, would be a good sign. Cells in the negative control—the floor—are not supposed to grow. But now, Sarah explains, she doesn't trust the controls because the experiments didn't grow either. She wonders out loud, "What if the controls aren't growing for the same reasons that the experiments with the lab's synthetic proteins aren't growing?" The "floor" had fallen out from under her.

In his classic book on replication, *Changing Order,* Harry Collins explains that replication, in experimentation, is a vital idea.[21] "Replicability," he writes, "in a manner of speaking, is the Supreme Court of the scientific system. . . . It corresponds to what the sociologist Robert Merton (1945) called the 'norm of universality.' Anybody, irrespective of who or what they are, in principle ought to be able to check for themselves through their own experiments that a scientific claim is valid."[22] No experiment, observes Collins, is convincingly confirmatory if it is an identical copy of a previous experiment. In order for a second try to be in some sense confirmatory, some differences have to be introduced between attempts.

> For an experiment to be a test of a previous result it must be neither exactly the same nor too different. Take a pair of experiments—one that gives rise to a new result and a subsequent test. If the second experiment is too like the first then it will not add any confirmatory information. The extreme case where every aspect of the second experiment is literally identical to the first is not even a separate experiment. Under these circumstances the second experiment would amount to no more than reading the first experimental report for a second time.[23]

The greater the difference between the initial success and its predecessors, the more confirmatory the latter attempts will be. Up to a point. If a replication attempt is confirmatory by means of "a skeptical fairground gypsy who had generated the confirmatory result by reading the entrails of a goat," then, says Collins, the confirmation is less convincing.[24]

Though, in theory, replication is a vital tenet of experimental methods, in practice, replication of the results of others is not as commonly attempted as one might expect.[25] The incentive structure of the experimental sciences has never encouraged it. Journals do not publish replication attempts, and departments don't give tenure on the basis of confirming someone else's neat result. Moreover, experiments are often, and increasingly, virtuosic performances, honed over many years of experience and practice with a particular set of techniques and requiring expensive equipment. In such circumstances, the question "Who is qualified to replicate?" becomes a decidedly nontrivial one. The barriers to replication grow, while replication itself becomes a test of very specific, cultivated skills that draw heavily from the tacit dimension. Thus, as Collins writes, "For the vast majority of science replicability is an axiom rather than a matter of practice."[26]

While replication of the results of others is rarely practiced for its own sake, within most labs, experiments are repeated frequently. In some cases, repetition is aimed at epistemic validation. As Betsy once explained to me, "Chemists do everything in triplicate." In many cases, however, repetition is a matter of pragmatic necessity, and yet carries epistemic weight. Confirmatory replication can perhaps be productively subsumed, then, into a whole host of different occurrences in laboratory life that leverage continuity and discontinuity toward knowledge ends, enriching or depleting the stockpile of confidence in results. Discontinuity can be achieved through changes of location, or people, or equipment. It can be planned or unplanned, controlled or uncontrolled, generative or destructive—producing doubt or confidence and, often, much frustration, felt in the pit of the stomach as much as in that proverbial center of rationality, the head.

In many cases, repetition is first and foremost a matter of pragmatic necessity. For example, practitioners replicate results from other labs when the ends of someone else's research can be made into means in one's own—that is, when a finding from another lab is recast as a building block or technique. This is because practitioners incorporate results of others' experiments not only to enlarge their store of knowledge but also to expand their technical proficiency and make new experimental things. A vast array of scientific findings has made its way from ends to means in this way.

Replication attempts that were undertaken in order to turn ends into means most often became topics of conversation when these replication attempts failed, creating roadblocks in research. On such occasions, suspicion about the original research took a long time to foment, as experimenters were much more likely to question the adequacy of their own attempt, checking and rechecking protocols and equipment. The threshold for successful repetition in such instances was, at first, measured by the result, a problem Collins has termed the "experimenter's regress." On one occasion, for example, Jenny, a student in Michael's lab, was trying to replicate a set of experiments from another lab in Arizona. Jenny had no interest in replication for its own sake. Rather, she wanted to build on these results and use them as a technique in her own research. She was therefore highly motivated to make the other lab's experiments work, ensuring a good-faith effort to reproduce the result, though also a fair amount of tinkering to achieve it.[27] Initially, Jenny spent three months tweaking and tinkering with the system, frustrated at her inability to replicate the original experiments. Her primary suspicion while attempting the experiments was that something was wrong with her equipment, which she checked and rechecked. But in the absence of certainty about the root cause of failure, she also began reluctantly to entertain the possibility that the original results were in some way faulty.

Finally, Jenny packed her bags and went to visit the lab that had originated the work. Neither the methods section of the original published paper, nor informal communication by phone and e-mail had provided Jenny with a clear recipe for success. Jenny was therefore heading off in order to mine for help in the tacit knowledge of those who had made these experiments work. The central thesis in Michael Polanyi's famous articulation of the concept is that "we know more than we can tell."[28] Polanyi explains, "An art which cannot be specified in detail cannot be transmitted by prescription, since no prescription for it exists. It can be passed on only by example from master to apprentice. This restricts the range of diffusion to that of personal contact."[29] Collins, who picked up the idea and extended its reach to his discussion of replication, has argued that acknowledging the tacit dimensions of experimentation replaces an "algorithmic model" of replication with an "enculturational" one.[30] Jenny had attempted the algorithm and, having failed, apprenticed herself to the experimental master. She stayed for a month, observing and participating in performing experiments she had failed to replicate at home. She returned confident that the experiments should be replicable. Yet she still couldn't get them to work on her own, until she finally found the culprit behind the failures: a faulty ultraviolet light. I tell this particular story not because it supports the primacy of the enculturational model—for all we know, with a working ultraviolet light, the algorithm would have done the job—but because the trip itself suggests that practitioners are well aware of the communicative limits of the methods section.

Another kind of opportunistic replication that is common in labs occurs within one and the same network of credit, and is a matter of some practical necessity, as when a new machine is purchased to replace an old one, a new reagent mixed, a new student shows up, and so on. The motive behind such repetition is not epistemic or technical confirmation. Nor is confirmation a good description of the spirit in which such repetition is carried out. Instead, the rationale behind such repetition is bluntly pragmatic: adjust to often-unavoidable change. Yet such opportunistic replication is nonetheless meaningful. It builds or depletes the storehouse of confidence in the results or in the stories lab members build around them; confidence, because a kernel of underdetermination persists around well-worked-out experimental systems whose opacity remains significant even after they have begun to yield publishable results.

This second kind of replication, in which repetition occurs by necessity, is most commonly spurred by an exogenous shock. Among these shocks, perhaps the most common catalyst is the simple movement of students and postdocs in and out of labs. Projects move hands. Graduate students go on to become postdocs, and postdocs (hopefully) find jobs. New graduate students and postdocs frequently inherit projects in different stages of development from their lab elders. Sometimes, both algorithm and culture are transmitted when outgoing and incoming lab members overlap. Frequently they are not.

To switch my ethnographic focus for a moment, a striking feature of Ron's account of his lab's projects was the influence of the movement of students on his sense of where projects stood. The process of transition from one lab member to another with respect to any particular project was a gradual one, Ron explained. First, new members were familiarized with lab techniques through work that would seldom become their primary focus in the lab. Lab members then most often picked up some piece of an existing project. Wei's Turing patterns project had been launched by a former graduate student in the lab; another postdoc was working on two projects that pre-dated her time in the lab—a protein toggle switch and a cancer project stemming from iGEM 2007. A newer graduate student had picked up the neuronal iGEM project from the preceding summer, and Meg was working on the diabetes project, which had originally been hatched for iGEM 2006.

As Ron recounted the state or fate of each project, it became clear that, when experiments changed hands, they didn't always continue to work, often for unknown reasons. Take, for example, the cell-cell communication project for mammalian cells. The project had been launched around 2004 by a graduate student, Malcolm, who had since received his doctorate and taken a position at another university. The project involved some cells sending signals and some receiving signals. Malcolm and Ron had devised an approach in which the two sides would be approached separately. Malcolm had managed to get the receiver side working very well. "He'd done it several times," explained Ron. On the sender side, Malcolm had shown the ability of mammalian cells to synthesize the desired protein. "It wasn't a lot," Ron recalled, "but it was enough to get me excited about it." But after Malcolm left, another graduate student in the lab took over the experiments and, so far, had been unable to make them work. In fact, the new graduate student on the project wasn't having any luck with either the receivers or the senders, even though he was in regular touch with Malcolm. Ron said:

> On the receiver side, I'm almost shocked that we can't replicate it. And on the sender side, I wasn't exactly sure whether it works or not, he hasn't been able to replicate that either, and so he's working on it.

"Hoyt"

After the meeting in the new building, Maria, Betsy, Michael, and I sit down for a quick catch-up in one of the common areas. Seated in one of dozens of brand-new chartreuse swiveling armchairs distributed throughout New Frick's common areas, Michael, with characteristic humor, explains the current state of anxiety in the lab. He scribbles three names of what I take to be proteins on a piece of paper. Lab members, he explains, had thought that CisN and CisD had rescued the auxotrophs. They were now hoping that it wasn't "Hoyt"—Hoyt!—meaning the previous lab building. The new built environment introduced so many changes at once that it opened up a marvelously paranoid space for conjecture. More than once, lab members run through the potential culprits. Was it the cells? Were they damaged? Did something thaw in the move? Something freeze that wasn't supposed to? One fear consistently rises to the top. What if Hoyt was contaminated and represented a special environment for growth? The auxotrophs, after all, are sickly bacteria that require extra nutrition from their environments to grow. What if Hoyt had somehow provided that nutrition? Betsy explained:

> The one thing that could have been the obvious problem would be the water. The issue would probably have been, because it isn't that we are getting extra growth, it's that we are getting *not* growth, the issue would be more that this water is purer than the water up there, and that there were maybe some small

amounts of contaminants up there that were helping it grow. For instance, Michael was thinking that maybe there was a problem in the pathway to an amino acid. So it's almost like a chicken and egg problem thing, where you need a little bit to even make the protein. So he was thinking there may have been a tiny bit of that in there, there was just enough to help it get started, and now if we don't even have that tiny bit, you know, I don't know.

Michael had another, more whimsical, theory that tied the *E. coli*'s experience to his own sense of dislocation. Second to a death in the family, he explained, moving is considered the most stressful activity. The *E. coli* in his lab that survived the key experiments experience millions of deaths in the family under *normal* circumstances. On top of that, they were suddenly moved. The comical-seeming projection onto some of the lab's smallest and most stressed contributors serves as a reminder that, though the spaces in which experimentation occurs have a kind of antiseptic impersonality that predicates objectivity on these spaces' erasure, for their human inhabitants—and sometimes, inadvertently, their nonhuman ones—they are very much places in which habits and attachments form.

We return to transparent buildings. In his essay on architectural transparency discussed earlier, Alloa's main concern is the significance of transparency—and its most hallowed conduit, glass—for twentieth-century cultural critics. Among the critics Alloa follows, Walter Benjamin stands out as a distinctly ambivalent critic of the material and its possibilities, as can be teased out of his famous and abundant meditations on "interiors," which include urban architectural reconfigurations of inside and outside, reflections on the nineteenth-century bourgeois household, and his attention to the psychological interiority so dominant in the cultural mediums of his day. Glass, for Benjamin, is a disrupter of any stable relationship between interiority and exteriority. He therefore extolls the revolutionary virtues of "moral exhibitionism" as an antidote to discretion, once an aristocratic virtue that had since "become more and more an affair of petit-bourgeois parvenus."[31] He likewise celebrates the impossibility of leaving traces on the slick and cold surface. If Brecht saw "Erase your traces!" as the motto of modern man, as Benjamin observes, one doesn't even have to bother erasing with glass. For Benjamin, glass is the "hard, smooth material to which nothing can be fixed. A cold and sober material, into the bargain. Objects made of glass have no 'aura.' Glass is, in general, the enemy of secrets."[32]

Famously, Benjamin's most incisive if fleeting ruminations on glass oppose transparency to the notion of "dwelling" and are contained in his compendium of notes on the Paris arcades, the glass-roofed streets that housed high-end shops and were largely destroyed during Hausmann's renovation of the city. Benjamin writes,

> The twentieth century, with its porosity and transparency, its tendency toward the well-lit and airy, has put an end to dwelling in the old sense. Set off against the doll house in the residence of the master builder Solness are the "homes for human beings." *Jugendstil* unsettled the world of the shell in a radical way. Today this world has disappeared entirely, and dwelling has diminished: for the living, through hotel rooms; for the dead, through crematoriums.[33]

Elsewhere, Benjamin writes,

> In the imprint of this turning point of the epoch, it is written that the knell has sounded for the dwelling in its old sense, dwelling in which security prevailed. Giedion, Mendelssohn, Le Corbusier have made the place of abode of men above all the transitory space of all the imaginable forces and waves of air and light. What is being prepared is found under the sign of transparency.[34]

In *The Architectural Uncanny*, Anthony Vidler includes transparency in a list of contemporary architectural themes that play on the "unhomely," a literal translation of the German word *unheimlich*, usually translated as "uncanny." The unhomely describes a sensibility that sees the familiar and comfortable invaded by an alien presence, where estrangement arises out of the distance between the self and the self as seen at a distance.[35] It is perhaps not surprising then that, in the move from Frick to Frick, it was Old Frick that became the source of suspicion, a building that had perhaps been on more intimate terms with the cells in the experiments than lab members had known, silently and imperceptibly nourishing auxotrophic bacteria. The building was old. It was worn. Once lived in but now unoccupied, and Gothic to boot, Old Frick was like a haunted house, "the most popular topos of the 19th century uncanny,"[36] while New Frick was the slick abode of openness, collaboration, and tracelessness.

In subsequent weeks, the auxotrophs rallied. Their recovery was as mysterious as their refusal to grow. No one quite knew what had happened, though

Betsy had a new theory, which was that a key reagent had gone bad. They had been using a fairly low concentration of the reagent in the first place. If it was old, and it had thawed, she explained, that may have been the final straw. When I asked whether, given that everyone had their own mix of the reagent made at different times, this was likely the problem across the board, she admitted that that was the one element that made the theory "iffy." Yet the theory helped Betsy restore her sense of security in work that consumed most of her days. She could move on.

Over time, and with a little distance, Michael began to include the story of the move in some of his talks. In lectures on the auxotrophs rescue project, Michael now explains that when the project first started to produce results, he was doubtful. When different members of his lab were able to replicate the experiments, his confidence grew, yet he still wasn't entirely sure. Then, Princeton University pitched into the effort by building an entire building just to test the replicability of his lab's results. Now he was convinced. Though obviously tongue-in-cheek, the irony of this version of events is most certainly lost on most of Michael's audience members, who cannot possibly know the extent to which the members of Michael's lab at the time had felt that the building had not been built for them.

Shortly after the return to semi-normalcy, a new grad student presented her work on the auxotrophs project. By new, I mean to the lab, not to the chemistry department. Mia had been a grad student with another group, but had absconded. Her presentation therefore detailed another occasion of routine replication of experiments: competence building. Mia was an organic chemist. She was learning to run the experiments in the Hecht lab by replicating auxotroph experiments and also testing a few new rescue contenders. Mia reported that she had managed to replicate experiments and grow auxotrophs with synthetic proteins again after the setbacks of the past month. Michael noted emphatically that things were back on track.

Halfway through her presentation, Mia reveals some new results. She has tested a new auxotroph that grows with a protein from the synthetic library. Michael then asks some fairly basic questions about the experiments: How long does it take for colonies to form? How many colonies were there? and so on. Then suddenly, he changes tack, asking after Mia's intuitive sense. Since the exchange unfolds in front of the entire lab, it has some elements of performance, a ritual renewal complete with an initiate, in the aftermath of some heavy doubt. To focus attention on intuition, Michael changes the tone of his voice and the manner in which he asks questions. He smiles. His expression becomes quizzical. He then invites Mia to share her gut feeling: Is the new auxotroph being rescued by the lab's synthetic protein? She answers with conviction. Yes, she is convinced by the results. And Michael makes sure to add that he is reassured that the results are real. Some intuition will have to do for now. These are exigencies of experimentation with biological artifacts that are not (yet) fully known or understood. As far as technoscience is concerned, both their transparency and opacity densely signify the contemporary epistemic predicament of experimentation with novel biological things.

Epilogue

This book is not intended as an overview of a new field. Nor does it seek to distill the broad potential cultural significance of the unwieldy set of research activities that have adopted the name synthetic biology. The book's scope is limited; ethnographically, its attention is fixed on the near at hand. Such scope and scale are recognizably anthropological. "Anthropologists," writes Rabinow, paraphrasing Clifford Geertz, are the "miniature painters of the social sciences, attuned to small events, minor processes, petty breakdowns, and the like as places to pause and inquire as to what is happening in front of us."[1] I have attempted to cultivate this scale of inquiry without steeping my analysis in empirical detail. The near requires just as much conceptual work as does the global, or the massive, or the remote.

In the introduction, I situated these choices of scope and sites in terms of the particular descriptive challenges presented by hype technoscience. Here, I would like to link these choices to broader problems of anthropological relevance that have hovered in the background of this study from its inception. As has been widely noted in the discipline, anthropology has been undergoing a prolonged crisis of relevance, sometimes expressed as a nostalgic craving for a more prominent role in public discourse that will restore to anthropology the glamour of a bygone era and also protect it against the budget cuts and rampant utilitarian chauvinism that threaten the humanities.[2] Yet so many of the paths to relevance risk reaffirming and reinscribing the configuration of values that has eroded the role of the university and contracted the exploratory space of academic work.

Studies of science and technology, all things considered, have been doing pretty well. I began this project, perhaps not surprisingly then, with the idea that relevance was baked into my study because of the world-changing potential of the technologies in question. But this particular way of framing relevance, which, in point of fact, takes the urgency of our subjects and objects as its measure, has a kind of parasitic structure that also further amplifies an existent ecology of relevance, urgency, and attention.[3] This, of course, isn't all bad. Not all objects of study raise the same sorts of difficulties. There are many urgent problems—from the impending global disaster of climate change to ever-increasing racialized social inequality—that are justifiably prioritized and to which anthropology has much to contribute. Yet the amplification also means the routinization of objectification and attention along familiar lines, which, when brought to bear on particular kinds of objects, does certain kinds of constructive labor.

It seems worth noting that this parasitic element has not always been the norm, though the alternatives had their own significant problems. Take Margaret Mead's classic, *Coming of Age in Samoa*, for example, which is itself a model work of public engagement by the scholar most often framed as the discipline's most successful public intellectual.[4] The book builds an explicit contrast between American and Samoan adolescence. The problem of adolescence, in the book, is framed as timely, but the Samoan subjects, to the book's and the field's detriment, are made to operate outside of history. That is, one of the sins of this earlier anthropology was that it framed subjects as removed from historical processes, wherein this very remove allowed these subjects to populate a natural laboratory for social theory. Today, on the contrary, we often seem to rely on the relevance of the activities in which our interlocutors are engaged to justify our own.

Such changes to the very structure of claims to relevance seem worth taking into account, and also thinking about, in relation to the concretion of values, demands, and temporalities that are susceptible to unconscious reproduction. It has become something of an anthropological prime directive to acknowledge that no observer ever steps outside the broader patterns of life, of which, if they are lucky or have worked hard enough, they can see glimmers. But there are moments when, through different kinds of cues, we suddenly sense that our performance of a cultural moment outpaces our ability to observe it, and the question becomes how to work *both within and a little against* the propellant, self-assembling machinery of the present.

In this book, I have taken advantage of some of the now quaint-seeming and untimely elements of lab ethnography, anthropology, and academic life that have more often than not been framed as liabilities in recent years— things like slowness, and the privileging of the local, as well as the absence of a clear path to utility.[5] I have used these various liabilities to highlight some thematic registers that cut across those that most vociferously announce themselves as native to contemporary technological frontiers.[6] Thus, for example, the promissory rhetoric insists that we ask what difference synthetic biology might make for The Future. In answering that question as posed, the anthropologist participates in the work of objectification, thereby bringing a particular future one tiny step closer to fruition. Attuning ourselves to the myriad problems of laboratory research in hype technoscience as it unfolds, as a counterweight, reminds us that the future is also the moment that follows directly after this one, in which something will have worked, or not, producing a tinge of excitement or a pang of disappointment; raising expectations or deflating them; eliciting a fresh thought or a sense of stagnancy; inviting a minor modification or suggesting an altogether new path, tactic, or question. My choice of highlighting this thematic range and scale latent within a biotechnological frontier isn't meant to stand by itself, replacing all others; it does not claim superior access or fealty to what's *really going on*, closing the gap between hubris and humility. It seeks complementarity, not domination. Its greatest potential lies in generating some productive tension.

Life Goes On

Much of the anthropology of the biosciences queries life's dissipating boundaries, teasing out an important implication of Michel Foucault's famous argument that "life" is the residue of the structured grid that has brought it into appearance, a residue subject to the vagaries of history and transformation.

And so, "life" may end.[7] I pick this line of argument up here, in the very last pages, in order to refract it through the problem of forging a position that is *both within and a little against* this constitutive discourse.

Today, at the frontiers of the biosciences, scientists, social scientists, and humanities scholars find the dissipating boundaries that perhaps portend life's final dissolution, an ironic twist for "the century of biology," which is where this book began. The Foucauldian framing resonates with the profusion of what Stefan Helmreich terms "limit biologies," areas of the biosciences that query life's boundaries and that include synthetic biology.[8] Indeed, concern with the meaning and limits of "life" is a common element of native discourse, discernible during my fieldwork. The engineering approach to synthetic biology practiced in Ron's lab toys with the view that life is a construction material like any other; Michael's lab uses their tiny proteins to argue that the primordial ingredients of life may not have been that special, and that life is easy "to do."

I have left questions about "life" largely to the side, first, because this book's thematic concerns lie elsewhere, and second, because I have been consistently drawn to a second-order question: What can we make of the profusion of discourses about and inquiries into the end of "life"? Asking this question turns one Foucauldian maneuver against another, making the first-order question, about how synthetic biology portends this end, harder to pose, for it draws anthropological handlings of the end of "life" into the same frame as the discourses and activities they are purportedly about.

Lévi-Strauss once argued that academics become interested in things just as they come to an end.[9] He was referring to the "primitive." In the early 1990s, Emily Martin borrowed Lévi-Strauss's observation to make sense of the explosion of interest in the body.[10] She argued that the body as a thing conceived and experienced in a way suited for Fordist mass production was coming to an end. Hers was not an "owl of Minerva" sort of argument. Rather, she suggested that the intensification of interest that accompanies endings was a matter of massive tectonic shifts in the economic base. Such intensifications of interest did not, therefore, suggest greater understanding of those things coming to an end. Quite the contrary. In missing the political economic transformation that leads to the body's demise, this heightened interest both fails to grasp the impending end and operates unaware of its own root cause.

But, one may ask, what about discourses that are themselves concerned with limits and endings? For the intensification of interest, in the case of "life," is not directed towards "life" itself. Rather, the interest is directed towards whether and how "life" ends. What does such an intensification signify? Could we read this in a way that ratchets Foucauldian thought even further, which suggests the possibility that what we should really pay attention to is the explosion of discourse about the end of "life" as the locus not of some real end or diminution, but rather the opposite, of some heightened tactical significance?

Melinda Cooper does something like this in a critique leveled at recent valorizations of "life."[11] Cooper observes that, in recent decades, efforts to query the limits of life have been deeply entangled with the search for sources of economic growth after the decline of profits from industrial modes of production in the 1960s and 1970s. She argues that, subsequently, pharmaceutical and petrochemical companies turned to biotechnology, initiating a speculative turn in which current economic value is calculated based on promises to overcome the limits of biology sometime in the future. Companies and countries now tout investments whose returns are based on the ability, not yet possessed, to transcend biological limits. Cooper thus argues that the interest in, and valorization of, the limits of "life" across a host of fields reflects the logic of "life" as capital under neoliberalism. Under this view, the historicization of "life," and the efforts to diagnose the practical and conceptual symptoms of its end, become, at least, a sign of the times, if not a full-fledged political economic instrument.

It is, of course, too late in this book to delve into theories of neoliberal biopolitics, nor would I be the right guide. I therefore submit an alternate reading of what this interest in the end of "life" does: namely, that discourses about grand conceptual endings toy with a kind of vertigo, stipulating the dissolution of that which seems indissolubly part of our conceptual grid. From this vantage point, we ponder, what is life without "life"? This is why hailing the imminent dissolution of "life" (or "man" or "the author" or "history") sometimes does work orthogonal to the diagnosis of the final dissolution of the conceptual grid that brought these shadows into being. Hailing the imminent dissolution of "life" brings some affective dimensions of "life" to their exalted peak. It is in this murky space between the final dissolution that would bring "life" to an end, and the gesturing toward that end that raises the affective load of "life" to a fever pitch, that the concern with life's limits operates across domains. We are rapt by the possibilities of another world, a "lifeless" world, just beyond the horizon of the present. Our very being rapt suggest that the end of "life" is part provocation, nigh (for now), while life goes on.[12]

How might we think about such affective dimensions that are attached to diagnosing key conceptual endings? I'll trade in Foucault for a moment for the historian of ideas Arthur Lovejoy. A student of William James and a contemporary of John Dewey, Lovejoy was the founder of the discipline of the history of ideas and a towering intellectual figure in his time, whose star markedly declined in subsequent decades. Lovejoy's obsolescence is already the subject of ridicule in Saul Bellow's novel *Herzog*, written in the mid-1960s, but it was Quentin Skinner's searing critique of Lovejoy's reification of ideas in a mostly subjectless universe that really did him in.[13] Yet Lovejoy has been defended of late, and put to use, in a disparate set of efforts. Here, I zoom in on one particular aspect of his thought. As I read him, Lovejoy equips us to confront affective qualities that attach themselves not to the substance of ideas, but rather to something more akin to their form. In this case, the formal quality I have in mind is "endings."

In his classic work, *The Great Chain of Being*, Lovejoy articulates a project for the history of ideas that draws on, but is separable from, the history of thought.[14] Lovejoy explains that, though it shares much of the same raw material with the history of thought, the history of ideas divides and sorts the material differently. He likens this project to the field of analytical chemistry. The history of ideas, in Lovejoy's hands, "cuts into" systems of thought, and breaks them up into "component elements," which Lovejoy calls "unit-ideas."[15] The term "unit-idea" is not, I confess, particularly lovable, and it has perhaps invited much of the misunderstanding that surrounds Lovejoy's work. Nevertheless, we might still benefit from some attention to what "unit-ideas" tell us about ideational objects. For one thing, they suggest that ideas are subunits of something else. Ideas are ways of breaking up philosophical doctrines that are taken to be "hard-and-fast."[16] The suggestion, then, is that "unit-ideas" provide points of incision for bringing to light groupings, patterns, and relations that cannot be apprehended when "doctrines" are viewed as totalities or wholes. As Lovejoy writes, "One of the results of the quest of the unit-ideas . . . is, I think, bound to be a livelier sense of the fact that most philosophic systems are original or distinctive rather in their patterns than in their components."[17] The usefulness of Lovejoy's approach, for my purposes here, lies both in the assortment of things that can be construed as unit-ideas, and in his emphasis on affect. As Lovejoy explains, the historian of ideas will seek to excavate the "common logical or pseudo-logical or *affective* ingredients behind the surface dissimilarities" that constitute what she encounters as doctrine or thought.[18]

What are these "unit-ideas" that are the proper object of investigation? Lovejoy does not attempt a formal definition, but rather, provides an oddly and whimsically informal numbered list of "heterogeneous," "recurrent dynamic units."[19] Here, I focus on the third item on Lovejoy's list: something he terms "metaphysical pathos." Lovejoy introduces the concept of metaphysical pathos as an attempt to name a "factor" having to do with certain historical "susceptibilities" that he finds to have been previously unnamed.[20] As he explains, "'Metaphysical pathos' is exemplified in any description of the nature of things, any characterization of the world to which one belongs, in terms which, like the words of a poem, awaken through their associations, and through a sort of empathy which they engender, a congenial mood or tone of feeling on the part of the philosopher or his readers." And he continues with some examples:

> Now of the metaphysical pathos there are a good many kinds. . . . There is, in the first place, the pathos of sheer obscurity, the loveliness of the incomprehensible, which has, I fear, stood many a philosopher in good stead with his public, even though he was innocent of intending any such effect. . . . The reader doesn't know exactly what they mean, but they have all the more on that account an air of sublimity; an agreeable feeling at once of awe and of exaltation comes over him as he contemplates thoughts of so immeasurable a profundity—their profundity being convincingly evidenced to him by the fact that he can see no bottom to them. Akin to this is the pathos of the esoteric. How exciting and how welcome is the sense of initiation into hidden mysteries! And how effectively have certain philosophers . . . satisfied the human craving for this experience, by representing the central insight of their philosophy as a thing to be reached, not through a consecutive progress of thought guided by the ordinary logic available to every man, but through a sudden leap whereby one rises to a plane of insight wholly different in its principles from the level of the mere understanding.[21]

What Lovejoy isolates here is the affective dimension of encountering thought that pleases or displeases, seduces or repels, not through the direct encounter with its "content," but through some seemingly secondary formal qualities. The "unit-idea" thus emerges as a very odd and intriguing sort of object.

If Lovejoy gives us the pathos of sheer obscurity and the pathos of the esoteric among others, I'll submit my own candidate: the pathos of vertiginous endings. To pay attention to this particular register of concern with the end of "life" is to render explicit some of the hold that certain narrative moves exert on thinking and theorizing that can themselves resonate with, or be appropriated toward, various ends.

Life goes on in other senses as well. By their principal investigators' estimations, the years that have elapsed since I conducted the fieldwork for this book have been productive ones in both Michael's and Ron's labs.

In a significant breakthrough, members of Michael's lab—some old, some new—managed to figure out how their de novo proteins work to rescue the auxotrophic bacteria. Two of their synthetic proteins alter gene regulation, causing overexpression of other enzymes in the cell; one of the lab's proteins is a bona fide enzyme.

Experimentally, the enzymatic activity took the form of some whitish extrusion in the petri dish. The first time Michael was shown this residue, he reports, he suggested that the graduate student running the experiments chuck it, thinking it signified a failed experiment. Luckily, he only half jokes, the student didn't listen. The new findings have worked retroactively to validate the original work that proved so hard to publish.

Meanwhile lab members have also come up with a method for making insulin that undercuts current pharmaceutical production costs by an order of magnitude. They are now figuring out how to bring their cheaper insulin to market, illustrating that the bioeconomy, for technoscientists, is never that far away. Michael's own interests now lie at the border of synthetic biology and astrobiology, where he queries possible life-forms and reactions using his lab's de novo proteins. Research- and morale-wise, the last few years have been good ones.

At MIT, Ron now directs the Center for Synthetic Biology, based in the Department of Bioengineering. In contrast to the "elite periphery" described in this book, a more central positioning would be hard to come by. Ron's lab has grown in size considerably, counting over two dozen researchers as its members. And, whereas at Princeton Ron's lab toiled in relative isolation, in Boston, Ron actively manages the number of collaborations in which his lab is involved so as to avoid being spread too thin. Research-wise, Ron's lab has been pursuing many new projects. Some highlights include projects related to cancer immunotherapy research and organ design. Lab members have also been figuring out how to wire circuits using RNA, instead of the transcriptional regulation that has served as the basis for the engineering approach to synthetic biology up to this point.

In Ron's view, the emphasis on systems-level biological engineering will likely be synthetic biology's most enduring legacy. He argues that if the term "synthetic biology" falls out of use, it will not be because the field failed, but because it became ubiquitous—its concepts, techniques, and tools integrated into laboratory practices around the world.

Notes

Introduction

1. J. Craig Venter and Daniel Cohen, "The Century of Biology," *NPQ: New Perspectives Quarterly* 21, no. 4 (2004): 73.

2. Ibid.

3. See, for example, Elizabeth Pennisi, "Synthetic Genome Brings New Life to Bacterium," *Science* 328 (2010): 958.

4. Paul Rabinow, *Essays in the Anthropology of Reason* (Princeton, NJ: Princeton University Press, 1996), 92.

5. Stefan Helmreich, *Alien Ocean: Anthropological Voyages in Microbial Seas* (Berkeley: University of California Press, 2009), 7.

6. For an elegantly crafted and relatively recent example, see D. Graham Burnett, *Trying Leviathan: The Nineteenth-Century New York Case That Put the Whale on Trial and Challenged the Order of Nature* (Princeton, NJ: Princeton University Press, 2007).

7. Throughout the book, I reference the last names of the heads of the two labs. Since I have changed the names of many of the other researchers who appear in the book to preserve their anonymity, I do not use last names when referring to other lab members.

8. A quick browser search for "genetic engineering kit" yields no shortage of results; many kits are marketed on the basis of their accessibility to total novices: "No experience needed."

9. Christopher M. Kelty, "Outlaw, Hackers, Victorian Amateurs: Diagnosing Public Participation in the Life Sciences Today," *Journal of Science Communication* 9, no. 1 (2010): C03.

10. International Center for Theoretical Sciences, January 24, 2013, "The Architecture of Biological Complexity—Sydney Brenner," https://www.youtube.com/watch?v=5uV73Vr0Os0.

11. Rabinow, *Essays*, 93. Emphasis in original.

12. Matthias Gross, *Ignorance and Surprise: Science, Society, and Ecological Design* (Cambridge, MA: MIT Press, 2010), 52.

13. Sophia Roosth, "Biobricks and Crocheted Coral: Dispatches from the Life Sciences in the Age of Fabrication," *Science in Context* 26, no. 1 (2013): 153–171.

14. Paul Voosen, "Synthetic Biology Comes Down to Earth," *Chronicle of Higher Education*, March 4, 2013, https://www.chronicle.com/article/Synthetic-Biology-Comes-Down/137587.

15. See Max Weber, "Science as a Vocation," in *From Max Weber: Essays in Sociology*, trans. and ed. H. H. Gerth and C. Wright Mills (New York: Oxford University Press, 1946), 129–156. Irony is perhaps built into this sort of question because, under some readings, Weber was opposed to the view that a vocation could be pursued on instrumental grounds. See Steven Shapin, *The Scientific Life: A Moral History of a Late Modern Vocation* (Chicago: University of Chicago Press, 2008). As David Owen and Tracy Strong write, "Vocational activity has as itself nothing of the instrumental; it is an end in itself (thus in some sense moral) but without reference to any grounding or act other than the freely chosen commitment of individuals to their own particular fates," in David Owen and Tracy B. Strong, "Introduction," in Max Weber, *The Vocation Lectures*, ed. David Owen and Tracy B. Strong (Indianapolis: Hackett, 2004), xiii. And, as many uses of the term would suggest, technoscience seems to require devotion to the instrumental. Yet there is sometimes more of Weber's characterization of the for-its-own-sake-ness of science as a vocation in technoscience than we might presume from our perusal of the synthetic biology marketing brochure that promises technological fixes and profit as the upshot of investment.

16. See Sophia Roosth, *Synthetic: How Life Got Made* (Chicago: University of Chicago Press, 2017), as well as Susanna Finlay, "Engineering Biology? Exploring Rhetoric, Practice, Constraints and Collaborations within a Synthetic Biology Research Centre," *Engineering Studies* 5, no. 1 (2013): 26–41.

17. Steven Shapin, *Never Pure: Historical Studies of Science as if It Was Produced by People with Bodies, Situated in Time, Space, Culture, Society, and Struggling for Credibility and Authority* (Baltimore, MD: Johns Hopkins University Press, 2010).

18. Niklas Luhmann, *Observations on Modernity* (Palo Alto, CA: Stanford University Press, 1998). A slightly expanded version of this observation and its implications can be found in my review of *Synthetic Aesthetics: Investigating Synthetic Biology's Designs on Life*, ed. Alexandra Daisy Ginsberg, Jane Calvert, Pablo Schyfter, Listair Elfick, and Drew Endy (Cambridge, MA: MIT Press, 2014); see Dan-Cohen, "Synthetic Biology in High Gloss," *BioSocieties* (2018).

19. Marilyn Strathern, "The Tyranny of Transparency," *British Educational Research Journal* 26, no. 3 (2000): 309–321 (312). Emphasis in original.

20. Ibid.

21. Michael Fischer, *Emergent Forms of Life and the Anthropological Voice* (Durham, NC: Duke University Press, 2003), 9.

22. Helmreich, *Alien Ocean*, 7.

23. George Marcus, *Ethnography through Thick and Thin* (Princeton, NJ: Princeton University Press, 1998).

24. Ibid., 45.

25. Matei Candea, "Arbitrary Locations: In Defense of the Bounded Field-Site," in *Multi-Sited Ethnography: Theory, Praxis and Locality in Contemporary Research*, ed. Mark-Anthony Falzon (New York: Ashgate, 2009), 27.

26. Andrew S. Balmer et al., *Synthetic Biology: A Sociology of Changing Practices* (New York: Palgrave Macmillan, 2016), 19–23.

27. Ibid., 21.

28. I invoke the relations between center and periphery to frame the ethnographic perspective cultivated in this book without the affect and politics usually attached to them in ethnographies that build a view (theoretical and ethnographic) from the margins. In anthropology, periphery represents the symbolic geography of the marginalized and the disempowered, as, for example, in Veena Das and Deborah Poole's volume, *Anthropology in the Margins of the State*, a collection whose collective view of the margins is indissoluble from a concern with violence. See Veena Das and Deborah Poole, "State and Its Margins: Comparative Ethnographies," in *Anthropology in the Margins of the State*, ed. Veena Das and Deborah Poole (Santa Fe, NM: School of American Research Press, 1991). Anthropologists have employed studies of the "periphery" in order to think through marginality and what it tells us about states, markets, regulatory regimes, and disciplinary technologies, not to mention the very boundary markers and power dynamics that produce margins in the first place. While I utilize the notion of margins or periphery to frame my study, my subjects are occasionally underdogs, but underdogs in an incredibly elite game. Their bare survival is not in question. They are not likely to be criminalized, or suffer material shortages. But, in a rather banal sense, they too are positioned in ways that implicate power relations, and intricately so. And since social life folds together many forms of institutional, organizational, and personal belonging, elements of "margins" and "periphery" organize intramural relations at high levels of privilege.

29. Laura Nader, "Up the Anthropologist—Perspectives Gained from Studying Up," in *Reinventing Anthropology*, ed. Dell Hymes (New York: Pantheon Books, 1972).

30. Candea, "Arbitrary Locations," 27.

31. Ibid., 35.

32. See, for example, Bruno Latour and Steve Woolgar, *Laboratory Life: The Construction of Scientific Facts* (Princeton, NJ: Princeton University Press, 1979), as well as Karin Knorr Cetina, *The Manufacture of Knowledge: An Essay on the Constructivist and Contextual Nature of Science* (New York: Pergamon Press, 1981). The focus of these laboratory ethnographies was most often some brand or branch of the sociology of knowledge, where specific laboratories and experimental systems could be made to stand for broader questions about what scientists do and how they do it. For this reason, many lab ethnographies, especially those that were properly sociological, submitted that the science to be observed was, in Thomas Kuhn's language, "normal science," or science pursued under

the auspices of a well-articulated paradigm, as opposed to "revolutionary science," which has become the more common focus of anthropological work in search of emergence. See Thomas Kuhn, *The Structure of Scientific Revolutions* (Chicago: University of Chicago Press, 2012 [1962]).

33. Annelise Riles, *The Network Inside Out* (Ann Arbor: University of Michigan Press, 2001).

34. See Paul Rabinow and Talia Dan-Cohen, *A Machine to Make a Future: Biotech Chronicles* (Princeton, NJ: Princeton University Press, 2005), which experiments with the chronicle form as a way of avoiding imposing a plotline on events whose shape and import were far too tentative for narrative closure.

35. Observers of synthetic biology have suggested a number of more robust ways of characterizing or organizing the field, despite its heterogeneity. For example, Pablo Schyfter argues that the "drive to make" is the uniting characteristic of the different approaches held together under the label "synthetic biology"; see Pablo Schyfter, "How a 'Drive to Make' Shapes Synthetic Biology," *Studies in History and Philosophy of Biological and Biomedical Sciences* 44, no. 4 (2013): 632–640. Schyfter's analysis works better for engineering-oriented synthetic biologists than for practitioners of some of the other approaches.

36. For a citational analysis that illustrates synthetic biology's heterogeneity, see Benjamin Raimbault, Jean-Philippe Cointet, and Pierre-Benoît Joly, "Mapping the Emergence of Synthetic Biology," *PLoS ONE* 11, no. 9 (2016): 1–19.

37. Michael Elowitz and Stanislas Leibler, "A Synthetic Oscillatory Network of Transcriptional Regulators," *Nature* 403, no. 6767 (2000): 335–338.

38. Timothy S. Gardner, Charles R. Cantor, and James J. Collins, "Construction of a Genetic Toggle Switch in Escherichia coli," *Nature* 403, no. 6767 (2000): 339–342.

39. Clemens Blümel has argued that the oscillator and the switch have been enrolled in persuasion practices where their function is largely a heuristic one. See Clemens Blümel, "Enrolling the Toggle Switch: Visionary Claims and the Capability of Modeling Objects in the Disciplinary Formation of Synthetic Biology," *NanoEthics* 10, no. 3 (2016): 269–287.

40. The first use of the term "synthetic biology" can be traced back to the work of Stéphane Leduc, who attempted to create artificial life in the early twentieth century. For a more complete account of synthetic biology's antecedents, see Luis Campos, "That Was the Synthetic Biology That Was," in *Synthetic Biology: The Technoscience and Its Societal Consequences*, ed. Markus Schmidt et al. (Dordrecht: Springer, 2010).

41. Jacques Monod and Francois Jacob, "Teleonomic Mechanisms in Cellular Metabolism, Growth, and Differentiation," *Cold Spring Harbor Symposia on Quantitative Biology* 26 (1961): 389–401.

42. Francois Jacob and Jacques Monod, "On the Regulation of Gene Activity," *Cold Spring Harbor Symposia on Quantitative Biology* 26 (1961): 193–211.

43. Thomas Kuhn, "Energy Conservation as an Example of Simultaneous Discovery," in *Critical Problems in the History of Science*, ed. Marshall Clagett (Madison: University of Wisconsin Press, 1959), 321.

44. This was not the end of the story, however. In 2016, Sanofi was reported to have produced no synthetic artemisinin and sold off its production site to a Hungarian company that would take over production. A temporary market glut of the wormwood-

derived stuff and a complicated tapestry of health, financial, and corporate considerations rendered the role of synthetic artemisinin in the global market fairly opaque and unstable. In this sense, what was billed as synthetic biology's first pharmaceutical bonanza has instead illustrated the difficulties involved in replacing, or interrupting, or even just complementing existing supply chains.

45. For more on Keasling, see Roosth, *Synthetic*, 61–64. Paul Rabinow was recruited to lead the "human practices" thrust at Synberc. Keasling figures prominently in some of Rabinow's subsequent books. See Paul Rabinow and Gaymon Bennett, *Designing Human Practices* (Chicago: University of Chicago Press, 2012). See also Paul Rabinow and Anthony Stavrianakis, *Demands of the Day: On the Logic of Anthropological Inquiry* (Chicago: University of Chicago Press, 2013).

46. Matthew Herper, "Hype in Genes," *Forbes*, July 23, 2007.

47. Peter Dizikes, "The Unraveling," *New York Times*, November 11, 2007, https://www.nytimes.com/2007/11/11/books/review/Dizikes-t.html.

48. Herper, "Hype in Genes."

49. Stephen Hilgartner, "Adventurer in Genome Science," *Science* 318, no. 5854 (2007): 1244–1245 (1244).

50. Chris Voigt, as quoted in Alexis Madrigal, "Synthetic Biology: It's Not What You Learned but What You Made," *Wired*, January 25, 2008.

51. Nicolas Wade, "Genetic Engineers Who Don't Just Tinker," *New York Times*, July 8, 2007.

52. Projects identified as synthetic biological are today pursued in 40 countries, in 700 organizations, funded by 530 funding agencies. See Yensi Flores Buesco and Mark Tangney, "Synthetic Biology in the Driving Seat of the Bioeconomy," *Trends in Biotechnology* 35, no. 5 (2017): 373–378.

53. Dominic Boyer, "Visiting Knowledge in Anthropology: An Introduction," *Ethnos* 70, no. 2 (2005): 141–148 (147).

1. Labs, Lives, Technoscience

1. Sharon Traweek, *Beamtimes and Lifetimes: The World of High Energy Physics* (Cambridge, MA: Harvard University Press, 1988).

2. Harold Abelson, Tom F. Knight, Gerald J. Sussman, et al., "Amorphous Computing Manifesto," https://groups.csail.mit.edu/mac/projects/amorphous/white-paper/amorph-new/amorph-new.html, 1996.

3. All unattributed interview quotations in the text come from personal interviews conducted between 2008 and 2011.

4. Oliver Morton, "Life, Reinvented," *Wired*, January 1, 2005, https://www.wired.com/2005/01/mit-3/.

5. For detailed examinations of proprietary regimes in synthetic biology, see Jane Calvert, "Ownership and Sharing in Synthetic Biology: A 'Diverse Ecology' of the Open and the Proprietary?" *BioSocieties* 7, no. 2 (2012): 169–187.

6. For a philosophical examination of synthesis in chemistry, see Stuart Rosenfeld and Nalini Bhushan, "Chemical Synthesis: Complexity, Similarity, Natural Kinds, and

the Evolution of a 'Logic,'" in *Of Minds and Molecules: New Philosophical Perspectives on Chemistry*, ed. Nalini Bhushan and Stuart Rosenfeld (Oxford: Oxford University Press, 2000), 187–207.

7. Natasha Myers's *Rendering Life Molecular: Models, Modelers, and Excitable Matter* (Durham, NC: Duke University Press, 2015) situates visualization within a broader set of embodied practices through which chemists get to know proteins.

8. Satwik Kamtekar, Jarad M. Schiffer, Huayu Xiong, Jennifer M. Babik, and Michael Hecht, "Protein Design by Binary Patterning of Polar and Nonpolar Amino Acids," *Science* 262, no. 5140 (1993): 1680–1685.

9. If Michael's lab's research speaks to deep evolutionary time, it also reaches toward outer space. Since the lab's research troubles understandings of the necessary ingredients for life, Michael had made some connections between his lab's research and the field of astrobiology, which queries the possibility for life on other planets. In this sense, Michael's lab's research fits into the rubric of what Stefan Helmreich terms *limit biologies*: "research projects in academic biology that explicitly query the limits of life, both as an empirical question and as a conceptual one"; see Stefan Helmreich, *Sounding the Limits of Life: Essays in the Anthropology of Biology and Beyond* (Princeton, NJ: Princeton University Press, 2016), xiii.

10. Michael Hecht, "Enabling Life with Molecular Parts Designed by Humans," Religion and Science Panel Discussion, November 8, 2010, Princeton University.

11. Evelyn Fox Keller, "What Does Synthetic Biology Have to Do with Biology?" *BioSocieties* 4, no. 2/3 (2009): 291–302 (293).

12. Ibid., 293. Latour's and Haraway's uses of "technoscience" are more complicated than they appear in Keller's quick gloss. Haraway credits Latour with the term's popularity, noting that the axis of distinction Latour attempted to undermine when he enrolled technoscience as a tactical intervention against prevailing approaches to the study of science was not the one that organizes a distinction between "pure" and "applied," but rather the one that separates "science" from "society." Latour used the label to set the heterogeneity of elements that can be brought into a sociotechnical network on an even plane. He explained, "I will use the word 'technoscience' from now on to describe all the elements tied to the scientific contents, no matter how dirty, unexpected or foreign they seem." Bruno Latour, *Science in Action* (Cambridge, MA: Harvard University Press, 1987), 174–175. Haraway's own usage draws out this heterogeneity as well, and emphasizes that the myriad players in technoscience exceed the boundaries of "science" and "engineering." Reflecting on her own use of the term, she writes, "I want to use technoscience to designate dense nodes of human and nonhuman actors that are brought into alliance by the material, social, and semiotic technologies through which what will count as nature and as matters of fact get constituted for—and by—many millions of people." Donna Haraway, *Modest_Witness@Second_Millennium.FemaleMan© _Meets_OncoMouse™* (New York: Routledge 1997), 50.

13. Constructivism itself can be broken down further into different programs and claims. For a discussion of the meanings of "social construction," see Ian Hacking, *The Social Construction of What?* (Cambridge, MA: Harvard University Press, 2000).

14. The resonance between this truism and the general turn against science in the public sphere has led to some fascinating soul-searching among science and technology

studies scholars worried that their own field played some part in creating the conditions for evacuating scientific authority. For a sampling of these concerns, see Sergio Sismondo, "Post-Truth?" *Social Studies of Science* 47, no. 1 (2017): 3–6, to which Harry Collins, Robert Evans, and Martin Weinel responded in "STS as Science or Politics," *Social Studies of Science* 47, no. 4 (2017): 580–586.

15. Keller, "What Does Synthetic Biology?," 294.

16. Ibid., 294.

17. For a thorough examination of the "epochal break thesis," see Alfred Nordmann, Hans Radder, and Gregor Schiemann, eds., *Science Transformed? Debating Claims of an Epochal Break* (Pittsburgh, PA: Pittsburgh University Press, 2011).

18. Alfred Nordmann, "The Age of Technoscience," in *Science Transformed? Debating Claims of an Epochal Break,* ed. Alfred Nordmann, Hans Radder, and Gregor Schiemann (Pittsburgh, PA: Pittsburgh University Press, 2011).

19. Nordmann's periodization might seem in tension with Rabinow's claim, cited in the introduction to this book, that the coupling of knowledge and power is one of the hallmarks of modernity. There is much reason to believe that intervention and utility have always been among the motives and aims of the modern sciences. Observing this fact furnishes Martin Carrier with his rebuttal to the epochal break thesis in an essay in the same volume as Nordmann's. See Carrier, "'Knowledge Is Power,' or How to Capture the Relationship between Science and Technoscience," in *Science Transformed?* 43–53. But insisting that knowledge has served the end of intervention in the sciences misses Nordmann's point. He acknowledges that utility has been a matter of concern, even a prime motivator. But the knowledge produced in its service still entailed forms of purification no longer available to today's technoscientists, who have let go of truth as a "regulative ideal." To be clear, Nordmann buys Latour's thesis that purification has always been futile. But he also argues that the ideal of purification has been available in a way that it is not today.

20. Bruno Latour, *We Have Never Been Modern* (Cambridge, MA: Harvard University Press, 1993).

21. Ibid., 25.

22. Ibid., 26.

23. Ibid., 25.

24. Evelyn Fox Keller, "Knowing as Making, Making as Knowing: The Many Lives of Synthetic Biology," *Biological Theory* 4, no. 4 (2009): 333–339.

25. David Edgerton, "'The Linear Model' Did Not Exist: Reflections on the History and Historiography of Science and Research in Industry in the Twentieth Century," in *The Science–Industry Nexus: History, Policy, Implications,* ed. Karl Grandin, Nina Wormbs, and Sven Widmalm (Sagamore Beach, MA: Science History Publications, 2004), 31–57.

26. Paul Forman, "The Primacy of Science in Modernity, of Technology in Postmodernity, and of Ideology in the History of Technology," *History and Technology* 23, no. 1/2 (2007): 1–152 (2).

27. Ibid., 2.

28. Ibid., 4.

29. Ibid., 53.

30. Ibid., 11.
31. Ibid., 5.

2. The Virtues of the Naïve View

1. Vanderbilt University, "Life Redesigned: The Emergence of Synthetic Biology," YouTube video, October 25, 2013, https://www.youtube.com/watch?v=fXdzHY7wJHQ.

2. "Naïve," in *Oxford Living Dictionaries*, https://en.oxforddictionaries.com/definition /naive.

3. See E. H. Gombrich, *Art and Illusion: A Study in the Psychology of Pictorial Representation* (Princeton, NJ: Princeton University Press, 2000).

4. See, for example, Michael Smithson, *Ignorance and Uncertainty* (New York: Springer-Verlag, 1989), and Robert N. Proctor and Londa Schiebinger, eds., *Agnotology: The Making and Unmaking of Ignorance* (Palo Alto, CA: Stanford University Press, 2008). In anthropology, see Roy Dilley and Thomas G. Kirsch, eds., *Regimes of Ignorance: Anthropological Perspectives on the Production and Reproduction of Non-knowledge* (New York: Berghahn Books, 2015). For a detailed review of the "ignorance" literature, see Matthias Gross, *Ignorance and Surprise: Science, Society, and Ecological Design* (Cambridge, MA: MIT Press, 2010).

5. Linsey McGoey, "The Logic of Strategic Ignorance," *British Journal of Sociology* 63, no. 3 (2012): 533–576.

6. *Keywords; For Further Consideration and Particularly Relevant to Academic Life* (Princeton, NJ: Princeton University Press 2018) makes a related intervention against the additive model: "The best interdisciplinary thinking arises not from a fusion but from a friction of methods. . . . Which is to say that interdisciplinary thinking is not more knowing than the knowledge given by the disciplines on which it depends (as it often claims to be), but less, and that is its value" (43).

7. Gaymon Bennett, "What Is a Part? Biofab Human Practices Report 1.0," 2010, http://biofab.synberc.org/sites/default/files/HPIP_DraftReport_Parts_1.0.pdf.

8. Paul Ricoeur famously observed that suspicion pervaded the writings of Marx, Freud, and Nietzsche. Paul Ricoeur, *Freud and Philosophy: An Essay on Interpretation*, translated by Denise Savage (New Haven: Yale University Press, 1970).

9. Rita Felski, "Critique and the Hermeneutics of Suspicion," *M/C Journal* 15, no. 1 (2012).

10. James Collins, "Bring in the Biologists," *Nature* 509, no. 7499 (2014): 155–156.

11. Ibid., 155.

12. Ibid.

13. Alfred Nordmann has observed that the limited time frame and inexperience of most of the practitioners are noteworthy features of the iGEM competition. Giving ignorance a central role in the competition, Nordmann observes, "the iGEM teams seek to find out through a strategic design process how much they can achieve with what little they know. They are not held back by seeking to learn all that would be needed for rationally engineering some biological structure or entity. Instead, they are invited and resolved to short-circuit the scruples of their teachers"; see Alfred Nordmann, "Synthetic

Biology at the Limits of Science," in *Synthetic Biology: Character and Impact*, ed. B. Giese, A. von Gleich, C. Pade, and H. Wigger (Berlin: Springer, 2014), 31–58.

14. Marilyn Strathern, "Cutting the Network," *Journal of the Royal Anthropological Institute* 2, no. 3 (1996): 517–535 (531).

15. Ibid., 524.

16. Ron's commitment to the primacy of abstraction relates to another interesting aspect of the philosophical attitude of engineers. When I once flat-footedly suggested that circuit abstractions are metaphors in biology, Ron immediately contested my claim, not on the grounds that these were literal representations, but on the grounds that these were instrumental abstractions in both biology *and* electrical engineering. The question was their utility, not their reality. In this sense, the instrumentalism of engineering perhaps predisposes some engineers against naïve realism, making them more sympathetic to constructivism and other forms of antirealism than scientists have tended to be.

17. See, for example, Michael Simpson et al., "Engineering in the Biological Sub-strate: Information Processing in Genetic Circuits," *Proceedings of the IEEE* 92, no. 5 (2004): 848–862.

18. In the introduction to their edited volume *Outsider Scientists: Routes to Innovation in Biology* (Chicago: University of Chicago Press, 2013), Oren Harman and Michael Dietrich identify a pattern in the kinds of interventions orchestrated by outsiders to the biological sciences. They note that many outsiders, especially those who possessed mathematical and physical skills, "sought to simplify matters in order, as it were, to see the forest independently of the trees" (16). In this sense, Ron and his engineer peers join theirs to a long line of interventions aimed at ordering the mess of biology using theoretical tools honed in other settings.

19. Yuri Lazebnik, "Can a Biologist Fix a Radio—Or, What I Learned while Studying Apoptosis," *Cell* 2, no. 3 (2002): 179–182.

20. Ibid., 180.

21. Annemarie Mol and John Law, "Complexities: An Introduction," in *Complexities: Social Studies of Knowledge Practices*, ed. John Law and Annemarie Mol (Durham, NC: Duke University Press, 2002), 1.

22. Ibid., 3.

23. Hirokazu Miyazaki and Annelise Riles, "Failure as an Endpoint," in *Global Assemblages: Technology, Politics, and Ethics as Anthropological Problems*, ed. Aihwa Ong and Stephen J. Collier (Malden, MA: Blackwell, 2005), 320–332.

24. Ibid., 327.

25. Sandra Mitchell, *Unsimple Truths: Science, Complexity, and Policy* (Chicago: University of Chicago Press, 2009).

26. Ibid., 13.

27. Ibid., 31.

28. Hans-Jörg Rheinberger, "Experimental Complexity in Biology: Some Epistemological and Historical Remarks," *Philosophy of Science* 64, no. 4 (1997): 246.

29. Ibid., 247.

30. Angela Creager et al., *Science without Laws: Model Systems, Cases, Exemplary Narratives* (Durham, NC: Duke University Press, 2007).

31. Hans-Jörg Rheinberger, "Gene Concepts: Fragments from the Perspective of Molecular Biology," in *The Concept of the Gene in Development and Evolution*, ed. Peter J. Beurton, Raphael Falk, and Hans-Jörg Rheinberger (Cambridge, UK: Cambridge University Press, 2008), 228.

32. Evelyn Fox Keller, *Refiguring Life: Metaphors of Twentieth-Century Biology* (New York: Columbia University Press, 1995), 95.

33. Evelyn Fox Keller, *The Century of the Gene* (Cambridge, MA: Harvard University Press, 2000).

34. Eric Lander and Robert Weinberg, "Genomics: Journey to the Center of Biology," *Science* 287, no. 5459 (2000): 1777.

35. For more on "the mapping paradigm," see Hans-Jörg Rheinberger and Jean-Paul Gaudillière, "Introduction," in *Classical Genetic Research and Its Legacy: The Mapping Cultures of Twentieth-century Genetics*, ed. Hans-Jörg Rheinberger and Jean-Paul Gaudillière (New York: Routledge, 2004), 1–5.

36. Erika Check Hayden, "Human Genome at Ten: Life Is Complicated," *Nature* 464, no. 7 (2010): 664–667.

37. An emphasis on the epistemological dimensions of complexity also might lead us to treat the contemporary alliance between complexity and vitalism with some caution. In recent work, Robert Mitchell notes that a vitalist turn is under way across many fields; see Mitchell, *Experimental Life: Vitalism in Romantic Science and Literature* (Baltimore, MD: Johns Hopkins University Press, 2013). The newly discovered complexity of biology has been a source of some thinly veiled schadenfreude among those social sciences and humanities scholars who have grown weary of reduction and simplification. Among these scholars, the discovery of complexity has served to bolster the view that "life is not a self-evident fact that can be taken for granted but rather a source of perplexity that demands new modes of conceptual and practical experimentation" (Mitchell, *Experimental Life*, 2). Yet we should be suspicious of the denunciation of simplification that celebrates complexity, warn Mol and Law, a position "so well established that it has become a morally comfortable place to be" (Mol and Law, "Complexities: An Introduction," 6). Thus, we might perhaps benefit from some circumspection when it comes to vitalist celebrations of biological complexity, not because simplicity will prevail, but because, in the process of celebrating, we weave together objects of knowledge and the unintended effects of the ways in which they have come to be known.

38. Evelyn Fox Keller, "What Does Synthetic Biology Have to Do with Biology?" *BioSocieties* 4, no. 2/3: 291–302.

39. Ibid., 295.

40. Ibid.

41. Drew Endy, as quoted in ibid., 296.

42. Ibid.

43. Ibid.

44. Ibid.

45. Maureen O'Malley, "Making Knowledge in Synthetic Biology: Design Meets Kludge," *Biological Theory* 4, no. 4 (2009): 378–389.

46. Jeffrey C. Way et al., "Integrating Biological Redesign: Where Synthetic Biology Came From and Where It Needs to Go," *Cell* 157, no. 1 (2014): 159.

47. George Church et al., "Realizing the Potential of Synthetic Biology," *Nature Reviews Molecular Cell Biology* 15, no. 4 (2014): 289.

48. Robert Proctor, "Agnotology," in Proctor and Schiebinger, *Agnotology*, 1–33.

49. Ibid., 2.

50. McGoey, "Logic of Strategic Ignorance," 554.

51. Paul Feyerabend, *Against Method* (New York: Verso, 2002 [1975]).

52. Ibid., 22.

53. Ibid., 113. Emphasis in original.

54. Ibid.

3. Looking for Patterns

1. Science studies scholars have documented many cases in which ambiguous results require that experimenters fix some definitions and thresholds. The classic and, to my knowledge, original exposition of this problem appears in Harry Collins's work on gravitational wave detectors; see Harry Collins, *Changing Order: Replication and Induction in Scientific Practice* (Chicago: University of Chicago Press, 1992 [1985]). For more recent ethnographic work on the topic, see Wolff-Michael Roth, "Making Classification (at) Work: Ordering Practices in Science," *Social Studies of Science* 35, no. 4 (2005): 581–621, as well as Wolff-Michael Roth and G. Michael Bowen, "'Creative Solutions' and 'Fibbing Results': Enculturation in Field Ecology," *Social Studies of Science* 31, no. 4 (2001): 533–536.

2. *The Princess Bride*, directed by Rob Reiner (Culver City, LA: Act III Productions, 1987).

3. Lorraine Daston, "On Scientific Observation," *Isis* 99 (2008): 97–110.

4. Ibid., 99.

5. This concept is a play on Margaret Lock's and Patricia Kaufert's "local biologies"; see Lock and Kaufert, "Menopause, Local Biologies, and Cultures of Aging," *American Journal of Human Biology* 13, no. 4 (2001): 494–504.

6. In "On Scientific Observation," Daston suggests that too much has been made of the "darkling character of perception, especially in the context of scientific observation." In her estimation, the twentieth-century interest in tacit knowledge has itself been somewhat to blame for the black-boxing of scientific observation. She writes, "It is perfectly possible to describe in considerable detail, as Fleck did, the stages by which perceptions coalesce into experience and above all to teach others to see in this way. The fact that a process cannot be reduced to a method or modelled by an algorithm or subjected to conscious introspection in all its aspects by no means implies that the process is irretrievably tacit, much less mystical, although that has indeed been the inference that much twentieth century philosophy of science drew" (101).

7. Alan Turing, "The Chemical Basis of Morphogenesis," *Philosophical Transactions of the Royal Society B* 237, no. 641 (1952): 5–72.

8. Conceptual underdetermination has received some attention from STS scholars, who, along with Rheinberger, have tended to argue that some fuzziness and underdetermination is, on the whole, salutary, and in a number of different ways. In the history

and philosophy of science, the gene has emerged as a kind of model system for this sort of conceptual ambiguity. Drawn into different traditions of instrumentation and experimentation, the gene's fuzzy boundaries have provided fodder for thinking about the way scientists proceed despite a lack of clear-cut, cross-disciplinary, or even intradisciplinary definitions. See, for example, Evelyn Fox Keller, *The Century of the Gene* (Cambridge, MA: Harvard University Press, 2000). Where broad, research-organizing concepts are concerned, ambiguity and underdetermination may allow for cooperation and communication among different communities of practice, while also furnishing experimenters with some open-endedness with which to mine the unknown. For the way ambiguity helps fuel cooperation between expert communities, see Peter Galison's discussion of trading zones in *Image and Logic: A Material Culture of Microphysics* (Chicago: University of Chicago Press, 1997); Susan Star and James Griesemer, "Institutional Ecology, 'Translations,' and Boundary Objects: Amateurs and Professionals in Berkeley's Museum of Vertebrate Zoology, 1907–39," *Social Studies of Science* 19, no. 3 (1989): 387–420. For a more recent examination of underdetermination in the sciences, see Lisa Messeri, "The Problem with Pluto: Conflicting Cosmologies and the Classification of Planets," *Social Studies of Science* 40, no. 2 (2009): 187–214. For a view of underdetermination's salutary effects, see Hans-Jörg Rheinberger, *An Epistemology of the Concrete: Twentieth-Century Histories of Life* (Durham, NC: Duke University Press, 2010).

9. Sigmund Freud, "Instincts and Their Vicissitudes," in *The Complete Psychological Works*, Standard Ed., vol. 14 (London: Hogarth Press, 1968), 117, quoted in Hans-Jörg Rheinberger, *Toward a History of Epistemic Things: Synthesizing Proteins in the Test Tube* (Stanford, CA: Stanford University Press, 1997), 12: "We have often heard it maintained that sciences should be built up on clear and sharply defined basic concepts. In actual fact no science, not even the most exact, begins with such definitions. The true beginning of scientific activity consists rather in describing phenomena and then in proceeding to group, classify and correlate them. Even at the stage of description it is not possible to avoid applying certain abstract ideas to the material in hand, ideas derived from somewhere or other but certainly not from the new observations alone. Such ideas—which will later become the basic concepts of the science—are still more indispensable as the material is further worked over. They must at first necessarily possess some degree of indefiniteness; there can be no question of any clear delimitation of their content. So long as they remain in this condition, we come to an understanding about their meaning by making repeated references to the material of observation from which they appear to have been derived, but upon which, in fact, they have been imposed. Thus, strictly speaking, they are in the nature of conventions—although everything depends on their not being arbitrarily chosen but determined by their having significant relations to the empirical material, relations that we seem to sense before we can clearly recognize and demonstrate them. It is only after more thorough investigation of the field of observation that we are able to formulate its basic scientific concepts with increased precision, and progressively so to modify them that they become serviceable and consistent over a wide area."

10. Rheinberger, *Toward a History*, 13. Emphasis in original.

11. The notion that concepts and ideas come from outside the lab furnishes an important part of Hannah Landecker's history of the concept of immortality, as it came

to apply to the longevity of cells. Landecker describes the technical means of achieving increased lifespan in cells, introduced by Alexis Carrel in the early twentieth century. Carrel famously cultured chicken heart cells that continued to beat in media for some hundred days. Carrel's use of the term immortality saturated laboratory practice with a set of colloquial notions. Yet these colloquial notions themselves would be changed by the adoption of immortality into scientific discourse. See Hannah Landecker, *Culturing Life: How Cells Became Technologies* (Cambridge, MA: Harvard University Press, 2007).

12. See Jeremy B. A. Green and James Sharpe, "Positional Information and Reaction-Diffusion: Two Big Ideas in Developmental Biology Combine," *Development* 142 (2015): 1203–1211.

13. James Scott, *Seeing Like a State: How Certain Schemes to Improve the Human Condition Have Failed* (New Haven, CT: Yale University Press, 1998).

14. Franz Boas, as quoted in E. H. Gombrich, *The Sense of Order: A Study in the Psychology of the Decorative Arts* (London: Phaidon Press, 2012 [1979]), 7.

15. Gombrich, *Sense of Order*, 7.

16. Ibid.

17. Tim Ingold and Elizabeth Hallam, "Creativity and Cultural Improvisation: An Introduction," in *Creativity and Cultural Improvisation*, ed. Elizabeth Hallam and Tim Ingold (Oxford: Berg, 2007), 1–24.

18. Ibid., 4.

19. The view laid out in this passage is fairly mysterious from the perspective of Ingold's other writings. In his essay, "Building, Dwelling, Living: How Animals and People Make Themselves at Home in the World," for example, Ingold rejects the notion that what distinguishes human-built environments from nonhuman ones is design, understood as the preconceptualization of a plan to be executed; see Tim Ingold, "Building, Dwelling, Living," in *The Perception of the Environment: Essays on Livelihood, Dwelling, and Skill* (London: Routledge, 2000). The passage from *Creativity and Cultural Improvisation* can therefore be interpreted as polemic. If polemic, it nonetheless usefully exaggerates a common perspective.

20. Ian Hacking, "The Self-Vindication of the Laboratory Sciences," in *Science as Practice and Culture*, ed. Andrew Pickering (Chicago: University of Chicago Press, 1992), 31.

21. The evolutionary point of view, too, turns out to be more complicated than one would perhaps guess. Visual patterns are thought to exist for many different reasons in the living world. Some patterns, like zebra stripes, are thought to achieve something called the "dazzle effect." That is, they make it difficult for predators to individuate animals in a herd and create optical illusions that make it more difficult for predators to pin down the direction of motion. This is not the story for giraffes, in which patterns are thought to serve the purpose primarily of camouflage. Yet giraffe patterns also overlay vascular patterns. Thus, patch patterns correspond to heat radiation patterns on the giraffe's body.

22. See Norwood Russell Hanson, *Patterns of Discovery: An Inquiry into the Conceptual Foundations of Science* (Cambridge, UK: Cambridge University Press, 1972).

23. See Thomas Kuhn, *The Structure of Scientific Revolutions* (Chicago: University of Chicago Press, 2012 [1962]).

24. Ludwik Fleck, *Genesis and Development of a Scientific Fact* (Chicago: University of Chicago Press, 1979).

25. Daston, "On Scientific Observation," 100.

26. E. H. Gombrich, *Art and Illusion* (Princeton, NJ: Princeton University Press, 2000 [1960]).

27. Ibid., 80.

28. Ibid., 81.

29. Ibid., 82. Christopher Wood explains that Gombrich repeatedly distanced himself from a radical constructionist reading of his work, and from figures like Nelson Goodman and Umberto Eco (see Christopher S. Wood, "Art History Reviewed VI: E. H. Gombrich's 'Art and Illusion: A Study in the Psychology of Pictorial Representation,'" *Burlington Magazine* [2009]: 836–839). And Gombrich does seem to suggest that European painters did in fact get better at representing appearances over time. Gombrich's book, *The Sense of Order*, more than the rest of his works, seems to suggest the supremacy of Western artistic forms as against "the primitive," a position that did not ingratiate him to the field of art history, which possessed a bent toward relativism and against the imperial eye of the art historian and his or her tastes. Nor did these views endear his efforts to the scholar who reviewed *The Sense of Order* for the *London Review of Books*, Clifford Geertz (see Geertz, "The Last Humanist," *London Review of Books,* September 26, 2002). Wood, however, highlights the independence of some of Gombrich's scholarship from Gombrich's own distaste for relativism.

30. The acceptance of this view helps unify observations about child art with other artistic traditions, notes Gombrich. Children, it has often been observed, draw from conceptual schemas. Their drawings conform to their knowledge of forms, rather than to the characteristics of an empirically observed reality; see Gombrich, *Art and Illusion*.

31. Gombrich, *Art and Illusion*, 90.

32. Ibid., 298. Nelson Goodman made much the same point when he wrote, "The myth of the innocent eye and the absolute given are unholy accomplices"; see Goodman, *Languages of Art: An Approach to a Theory of Symbols* (Indianapolis, IN: Hackett, 1976), 8.

33. Goodman, *Languages of Art*, 8. One of Gombrich's main adversaries in his account of convention was the preeminent nineteenth-century art critic John Ruskin, for whom the technical task of painting was a recovery of what he called the *innocence of the eye*. "That is to say," Ruskin explains, "of a sort of childish perception of . . . flat stains of colour, merely as such, without consciousness of what they signify, as a blind man would see them if suddenly gifted with sight." Ruskin, as quoted in Gombrich, *Art and Illusion*, 296. Ruskin, notes Gombrich, thus anticipated the doctrine of the Impressionists.

4. To the Editor

1. For a general discussion of authorship in the sciences, see Mario Biagioli and Peter Galison, eds., *Scientific Authorship: Credit and Intellectual Property in Science* (New York: Routledge, 2003).

2. Steven Shapin, *A Social History of Truth: Civility and Science in Seventeenth-Century England* (Chicago: University of Chicago Press, 1994).

3. Lorraine Daston, "Objectivity and the Escape from Perspective," *Social Studies of Science* 22, no. 4 (1992): 597–618.

4. Aileen Fyfe, "Peer Review: Not as Old as You Might Think," *Times Higher Education*, June 25, 2015, https://www.timeshighereducation.com/features/peer-review-not-old-you-might-think.

5. The different mechanisms were also tied to different claims about the veracity and reliability of published research. The Royal Society's *Philosophical Investigations* began each issue with a disclaimer that credit or blame for its communications lay with each contribution's authors. The society's aims were "importance and singularity." In contrast, the Academie Royale des Sciences in fact attempted to replicate research findings from the late 1700s through the 1830s, when the practice was abandoned. See Fyfe, "Peer Review."

6. Alex Csiszar, "Peer Review: Troubled from the Start," *Nature* 532 (2015): 306–308.

7. Ibid., 308.

8. Ibid.

9. Alex Csiszar, *The Scientific Journal: Authorship and the Politics of Knowledge in the Nineteenth Century* (Chicago: University of Chicago Press, 2018), 151.

10. Ibid., 152.

11. Csiszar, "Peer Review," 308.

12. Csiszar, *Scientific Journal*, 151.

13. Csiszar, "Peer Review," 308.

14. Ibid.

15. Fyfe, "Peer Review."

16. Csiszar, "Peer Review," 308.

17. Richard Smith, "Peer Review: A Flawed Process at the Heart of Science and Journals," *Journal of the Royal Society of Medicine* 99, no. 4 (2006): 178.

18. Sydney Brenner, "Moron Peer Review," *Current Biology* 9, no. 2 (1999): 1–9.

19. Ibid., 9.

20. The sociological interest in peer review arose largely in response to work by Harriet Zuckerman and Robert Merton; see Harriet Zuckerman and Robert K. Merton, "Patterns of Evaluation in Science: Institutionalization, Structure, and Functions of Referee Systems," *Minerva* 9 (1971): 66–100. For a recent example, see Michele Lamont, *How Professors Think* (Cambridge, MA: Harvard University Press, 2009), which examines peer review as an evaluative culture. Stefan Hirschauer notes that peer review is often viewed as objective by authors when judgments work in their favor, and as social and biased when they do not. In other words, he notes bias about bias. See Stefan Hirschauer, "Editorial Judgments: A Praxeology of 'Voting' in Peer Review," *Social Studies of Science* 40, no. 1 (2010): 71–103. There seem to me to be ample grounds for contesting this generalization along the lines that, despite their frequent irrationality, subjects, even academic ones, can on occasion show some consistency. But since, in this chapter, I track a case in which the editorial decision is a negative one, the point is somewhat moot.

21. Hans-Jörg Rheinberger, "'Discourses of Circumstance': A Note on the Author in Science," in *Scientific Authorship: Credit and Intellectual Property in Science*, ed. Mario Biagioli and Peter Galison (New York: Routledge, 2003), 309–323.

22. Emphasis added.

23. Shapin, *Social History of Truth*, 6.

24. Ibid., 7. Russell Hardin argues that trust can be explained entirely in terms of rational expectations without recourse to morality. He reserves the term "trustworthiness" for the morally encumbered variant of trust; see Russell Hardin, "Trustworthiness," *Ethics* 107, no. 1 (1996): 26–42. Since I use trustworthiness more colloquially in this chapter, I have chosen to stick with Shapin's terminology.

25. The literature on trust is vast. Georg Simmel's work on the subject has been highly influential; see George Simmel, *The Sociology of Georg Simmel*, ed. and trans. K. H. Wolff (New York: Free Press, 1950). Trust has proven a major concept for the elaboration of the notion of reflexive modernization. See, for example, Anthony Giddens, *The Consequences of Modernity* (Cambridge, UK: Polity Press, 1990). Karen Cook, Russell Hardin, and Margaret Levi organized a broad project on trust that resulted in a number of publications, including Karen S. Cook, ed., *Trust in Society* (New York: Russell Sage Foundation, 2001); Russell Hardin, *Trust and Trustworthiness* (New York: Russell Sage Foundation, 2004); Roderick Kramer and Karen Cook, *Trust and Distrust in Organizations: Dilemmas and Approaches* (New York: Russell Sage Foundation, 2004). In philosophy, see Paul Faulkner and Thomas Simpson, eds., *The Philosophy of Trust* (Oxford: Oxford University Press, 2017).

26. Some of the literature on trust and science focuses on relations of trust between scientists and publics. Some examples include Kyle Powys Whyte and Robert P. Crease, "Trust, Expertise, and the Philosophy of Science," *Synthese* 177, no. 3 (2010): 411–425, and Heidi Grasswick, "Scientific and Lay Communities: Earning Epistemic Trust through Knowledge Sharing," *Synthese* 177, no. 3 (2010): 387–409.

27. Shapin, *Social History of Truth*, 16.

28. Ibid., 19. Emphasis in original.

29. Steven Shapin and Simon Schaffer, *Leviathan and the Air-Pump: Hobbes, Boyle, and the Experimental Life* (Princeton, NJ: Princeton University Press, 2011).

30. Ibid., 58.

31. Ibid., 65.

32. Ibid. Emphasis in original.

33. Ibid.

34. Ibid., 66.

35. Ibid.

36. Donna Haraway, *Modest_Witness@Second_Millenium.FemaleMan©_Meets_OncoMouse™: Feminism and Technoscience* (New York: Routledge, 1997).

37. Ibid., 23. Haraway argues that such masculine modesty, an innovation when compared with the heroic masculinity that preceded it, did more than exclude certain groups from becoming viable witnesses; it structured the very differences that came to legitimate who can and cannot witness.

38. See Rick Kennedy, *A History of Reasonableness: Testimony and Authority in the Art of Thinking* (Rochester, NY: University of Rochester Press, 2004). See also Lorraine Daston, "The Moral Economy of Science," *Osiris* 10 (1995): 14.

39. Shapin, *Social History of Truth*, 412.

40. Ibid., 411.

41. Niklas Luhmann, *Trust and Power* (Cambridge, UK: Polity Press, 2017), 55.

42. Shapin, *Social History of Truth*, 412. Emphasis in original.

43. Ibid., 414.

44. Steven Shapin, *The Scientific Life: A Moral History of a Late Modern Vocation* (Chicago: University of Chicago Press, 2008). Other scholars have taken up this line of argument as well. For example, concurring with Shapin, and extending the empirical analysis of trust beyond science, Adam Hedgecoe has shown the centrality of personal trust in the regulation of scientific experimentation; Adam Hedgecoe "Trust and Regulatory Organisations: The Role of Local Knowledge and Facework in Research Ethics Review," *Social Studies of Science* 42, no. 5 (2012): 662–683.

45. Paul Rabinow, *The Accompaniment: Assembling the Contemporary* (Chicago: University of Chicago Press, 2011), 171.

46. The rhetoric of democratization that has often accompanied synthetic biology—whether through the celebration of the do-it-yourself community attached to it, or through the adoption of an open-source approach to biological parts—addresses these concerns not by grappling with them but by attempting to reconfigure precisely the boundaries between technoscience and public. See Sara Giordano, "New Democratic Sciences, Ethics, and Proper Publics," *Science, Technology and Human Values* 43, no. 3 (2018): 401–430.

47. Kennedy, *History of Reasonableness*.

48. Kennedy's book seems prescient in this regard. In the age of "fake news," the epistemological and educational causes of the crisis of credibility and the extent to which such a crisis might be related to an emphasis on critical thinking pose interesting questions.

49. Smith, "Peer Review," 178.

50. Experimental artifacts figure prominently in several works that challenge rationalistic accounts of scientific method. See, for example, Bruno Latour and Steve Woolgar, *Laboratory Life: The Construction of Scientific Facts* (Princeton, NJ: Princeton University Press, 1986). See also Nicolas Rasmussen, "Facts, Artifacts, and Mesosomes: Practicing Epistemology with the Electron Microscope," *Studies in the History and Philosophy of Science* 24, no. 2 (1993): 227–265.

51. Marcel Weber, *Philosophy of Experimental Biology* (Cambridge, UK: Cambridge University Press, 2005).

52. Ibid., 287.

53. See Robert Mitchell, *Experimental Life: Vitalism in Romantic Science and Literature* (Baltimore, MD: Johns Hopkins University Press, 2013), 22. Emphasis in original.

54. Lorraine Daston, "On Scientific Observation," *Isis* 99 (2008): 99.

55. Gilbert Ryle, *On Thinking* (Oxford: Basil Blackwell, 1979), 91.

56. Daston, "On Scientific Observation," 99.

5. On the Move

1. Peter Galison, "Buildings and the Subject of Science," in *The Architecture of Science*, ed. Peter Galison and Emily Thompson (Cambridge, MA: MIT Press, 1999), 3.

2. Thomas F. Gieryn, "Two Faces on Science: Building Identities for Molecular Biology and Biotechnology," in *The Architecture of Science*, ed. Peter Galison and Emily Thompson (Cambridge, MA: MIT Press, 1999), 423.

3. Alexander Leitch, *A Princeton Companion* (Princeton, NJ: Princeton University Press, 1978).

4. The most impressive structure to be built on the Princeton campus during my years as a graduate student there was the Lewis Science Library, designed by Frank Gehry. The building was a source of some general buzz, not only because of Gehry's status as a "starchitect," but also because of his recent legal squabbles with MIT over his design for that campus, the Stata Center, which turned out to be permeable to the elements and therefore supremely costly to maintain. The Lewis Center, with its bold steel planes, sharp angles, and cheery interior colors, opened to generally good reviews. It swiftly became one of my favorite places to work, owing to its manageable size and just-beyond-kindergarten-appropriate furniture, which contrasted heavily with the dungeonous depths of Princeton's main library, appropriately called Firestone. Firestone feels like a serious place. Lewis Library, in contrast, had been built with an eye for play.

5. Many scholars today identify much that flies under the banner of "transparency" with neoliberalism. In his manifesto-styled book on transparency, for example, the philosopher Byung-Chul Han situates transparency at the heart of our neoliberal age; see Byung-Chul Han, *The Transparency Society* (Palo Alto, CA: Stanford Briefs, 2015). Declaiming various meanings of neoliberalism, Hugh Gusterson notes that neoliberalism is sometimes synonymous with our "fishbowl society," characterized by ubiquitous metrics and audit practices; see Gusterson, "Homework: Toward a Critical Ethnography of the University," *American Ethnologist* 44, no. 3 (2017): 435–450. Gusterson frames universities as excellent places to study the spread and development of fishbowl neoliberalism. Other scholars have recently resisted a too-hermetic identification between transparency and neoliberalism. Contra Han, the historian of postwar France Stefanos Geroulanos argues that the reading of transparency as inherently neoliberal is too totalizing and ahistorical. Transparency, for him, is a myth that traverses modernity. It may have been opportunistically adopted by neoliberals or neoliberalism, but this is among the things one can say about it, not the only thing; see Stefanos Geroulanos, *Transparency in Postwar France* (Palo Alto, CA: Stanford University Press, 2017). Moreover, transparency has different histories in different contexts. For example, Geroulanos identifies a multifront revolt against "transparency" in mid-twentieth-century France that he argues did not take place in the same way in the United States. In some parts of the US university, "transparency" seems bluntly overdetermined by religious themes, modern and Enlightenment ideals, neoliberal tendencies, and corporate cultural adoptions, even creating some awkwardness where these different orders interact.

6. Emanuel Alloa, "Architecture of Transparency," *RES: Anthropology and Aesthetics* 53/54 (2008): 321–330.

7. Robinson Meyer, "How Gothic Architecture Took Over the American College Campus," *The Atlantic*, September 11, 2013, https://www.theatlantic.com/education/archive/2013/09/how-gothic-architecture-took-over-the-american-college-campus/279287/.

8. Johanna G. Seasonwein, *Princeton and the Gothic Revival, 1870–1930* (Princeton, NJ: Princeton University Art Museum/Princeton University Press, 2012), 22.

9. Erwin Panofsky, *Gothic Architecture and Scholasticism* (New York: Meridian, 1957), 43, as cited in Alloa, "Architecture of Transparency," 322.

10. Ruth Stevens, "Elements of New Frick Lab Join to Create 'Best Infrastructure' for Chemistry," September 2, 2010, https://www.princeton.edu/news/2010/09/02/elements -new-frick-lab-join-create-best-infrastructure-chemistry.

11. Ibid.

12. For a history of long-term research contracts between single corporations and universities, as well as broader discussions of the "university-industrial complex" in biotechnology, see Martin Kenney, *Biotechnology: The University-Industrial Complex* (New Haven, CT: Yale University Press, 1986).

13. Catherine Zandonella, "Alimta: Fundamental Science, Fundamental Benefit," November 18, 2011, http://research.princeton.edu/news/features/a/index.xml?id=6187.

14. See Ben Butkus, "Lilly and Princeton File Third IP-Infringement Suit against Generic Makers for Alimta Patent," Genome Web, May 6, 2009.

15. See https://www.princeton.edu/meet-princeton. This page may not have existed in this form when the patent battles were waging, but it is exemplary of research university rhetoric nonetheless.

16. The relationship with Eli Lilly also engendered some increasingly tense relations with neighbors in the Princeton Township, who, in a lawsuit, questioned Princeton's tax-exempt status, given that the school was accruing a sizeable fortune from Alimta royalties.

17. Marilyn Strathern, "The Tyranny of Transparency," *British Educational Research Journal* 26 no. 3 (2000): 309.

18. Transparent workplaces in the university are by no means limited to the sciences. They are ubiquitous. Walk down the halls of countless academic departments and you will find square little glass window holes in the doors or, in newer buildings, glass panels along the floors or walls. These little windows into offices produce tensions around valued privacy in the workplace. Minute acts of resistance to these conduits of visibility are routinely practiced by academicians. The little windows are covered with posters, art, and, most frequently, publications, which, when read critically, suggest that the scrutiny of the university belongs not in the conduct of individuals, but in their intellectual products.

19. Gusterson, "Homework." Gusterson argues that sociologists and fiction authors have seized on opportunities to observe and write about universities in recent years. Interestingly, this runs against the argument of an article by Aida Edemariam published in *The Guardian* ("Who's Afraid of the Campus Novel?" October 1, 2004), which makes the case that the campus novel has seen its heyday and is on the decline for precisely the reasons that Gusterson argues it is thriving. Edemariam quotes authors who argue that the campus has gone from being a humorous setup to being a tragic one, effectively doing in the once-thriving genre.

20. James Collins Jr., "The Design Process for the Human Workplace," in *The Architecture of Science*, ed. Peter Galison and Emily Thompson (Cambridge, MA: MIT Press, 1999), 402.

21. Harry Collins, *Changing Order: Replication and Induction in Scientific Practice* (Chicago: University of Chicago Press, 1992 [1985]).

22. Ibid., 19.

23. Ibid., 34.

24. Ibid., 35.

25. Ibid., 18. See also Ayelet Shavit and Aaron M. Ellison, "Toward a Taxonomy of Scientific Replication," *Stepping in the Same River Twice: Replication in Biological Research*, ed. Ayelet Shavit and Aaron M. Ellison (New Haven, CT: Yale University Press, 2017), 3–22.

26. Collins, *Changing Order*, 19.

27. See Karin Knorr Cetina, "Tinkering toward Success: Prelude to a Theory of Scientific Practice," *Theory and Society* 8, no. 3 (1979): 347–376.

28. Michael Polanyi, *The Tacit Dimension* (Chicago: University of Chicago Press, 2009). Crucially for Polanyi, as Park Doing points out in his review of tacit knowledge, "that 'we know more than we can tell' means that scientific progress cannot be controlled from the outside." See Park Doing, "Review Essay: Tacit Knowledge: Discovery by or Topic for Science Studies?" *Social Studies of Science* 41, no. 2 (2011): 301–306 (302).

29. Michael Polanyi, *Personal Knowledge* (London: Routledge/Kegan Paul, 1958), 52–53, as cited in Collins, *Changing Order*, 77. For Polanyi, "tacit knowledge" was a way of responding to the Soviet disdain for pure science and its attempt to put science under state control.

30. Collins, *Changing Order*, 57. For further elaborations of, and distinctions within, the tacit dimension, see Harry Collins, *Tacit and Explicit Knowledge* (Chicago: University of Chicago Press, 2010).

31. Walter Benjamin, "Surrealism: The Last Snapshot of the European Intelligentsia" [1929], in *Selected Writings, 1927–1930*, ed. Michael W. Jennings, Howard Eiland, and Gary Smith, vol. 2, part 1 (Cambridge, MA: Belknap Press of Harvard University Press, 1992), 206–221; cited in Alloa, "Architecture of Transparency," 327.

32. Walter Benjamin, "Experience and Poverty" in *Selected Writing, 1931–1934*, vol. 2, part 2, ed. Michael W. Jennings, Howard Eiland, and Gary Smith (Cambridge, MA: Belknap Press of Harvard University Press, 1999), 734.

33. Walter Benjamin, *The Arcades Project*, trans. Howard Eiland and Kevin McLaughlin (Cambridge, MA: Harvard University Press, 2002), 221.

34. Walter Benjamin, cited in Anthony Vidler, *The Architectural Uncanny: Essays in the Modern Unhomely* (Cambridge, MA: MIT Press, 1992), 217.

35. Vidler, *Architectural Uncanny*, 4.

36. Ibid.

Epilogue

1. Paul Rabinow, *The Accompaniment: Assembling the Contemporary* (Chicago: University of Chicago Press, 2011), 42.

2. Matti Bunzl, "The Quest for Anthropological Relevance: Borgesian Maps and Epistemological Pitfalls," *American Anthropologist* 110 (2008): 53–60.

3. For an examination of how a crisis chasing anthropology risks amplifying and augmenting existing logics, see Heath Cabot, "The Business of Anthropology and the European Refugee Crisis," *American Ethnologist* 46, no. 3 (2019): 261–275.

4. Margaret Mead, *Coming of Age in Samoa* (New York: Perennial Classics 2001 [1928]).

5. It goes without saying that such untimeliness can be engaged in only from a position of considerable privilege, as it runs against the pragmatic utilitarianism that Paul Forman so convincingly names as the normative principle of so much that goes on in the Euro-American university today.

6. There has been an assortment of interventions against the demands for speed in anthropology of late. See for example Vincanne Adams, Nancy Burke, and Ian Whitmarsh, "Slow Research: Thoughts for a Movement in Global Health," *Medical Anthropology* 33, no. 3 (2013): 179–197. See also Carlo Caduff, "Speed Crash Course," *Cultural Anthropology* 32, no. 1 (2017): 12–20. Caduff's piece is part of a special issue on speed that appeared in the journal *Cultural Anthropology* 32, no. 1 (2017), edited by Vincent Duclos, Tomás Sánchez Criado, and Vinh-Kim Nguyen.

7. See Michel Foucault, *The Order of Things* (New York: Vintage Books, 1994).

8. See Stefan Helmreich, "What Was Life? Answers from Three Limit Biologies," *Critical Inquiry* 37, no. 4 (2011): 671–696.

9. Claude Lévi-Strauss, *Tristes Tropique: An Anthropological Study of Primitive Societies in Brazil*, trans. John Russell (New York: Atheneum, 1967).

10. Emily Martin, "The End of the Body?" *American Ethnologist* 19, no. 1 (1992): 121–140.

11. Melinda Cooper, *Life as Surplus: Biotechnology and Capitalism in the Neoliberal Era* (Seattle: University of Washington Press, 2008).

12. What does this focus on grand conceptual transformation do in anthropology specifically? One answer might be that it serves as a homologue to a classical anthropological maneuver by translating the difference necessary for a certain kind of anthropological estrangement across the threshold of time, rather than cultural otherness. See Tobias Rees, *After Ethnos* (Durham, NC: Duke University Press, 2018). In this sense, a new exoticism is rung out of studies that replace "the encounter with a cultural other . . . with first contact with an unknown future." Nicolas Langlitz, *Neuropsychedelia: The Revival of Hallucinogen Research since the Decade of the Brain* (Berkeley: University of California Press, 2012), 240. In place of both historicism and culturalism, we have futurism.

13. Saul Bellow, *Herzog* (New York: Penguin, 1992), 209; Quentin Skinner, "Meaning and Understanding in the History of Ideas," *History and Theory* 8, no. 1 (1969): 3–53.

14. Arthur Lovejoy, *The Great Chain of Being* (Cambridge, MA: Harvard University Press, 1936).

15. Ibid., 3.

16. Ibid.

17. Ibid.

18. Ibid., 4. Emphasis added.

19. Ibid., 7.

20. Ibid.

21. Ibid.

Bibliography

Abelson, Harold, Tom F. Knight, Gerald J. Sussman, et al. "Amorphous Computing Manifesto." https://groups.csail.mit.edu/mac/projects/amorphous/white-paper/amorph -new/amorph-new.html, 1996.

Adams, Vincanne, Nancy Burke, and Ian Whitmarsh. "Slow Research: Thoughts for a Movement in Global Health." *Medical Anthropology* 33, no. 3 (2013): 179–197.

Alloa, Emanuel. "Architecture of Transparency." *RES: Anthropology and Aesthetics* 53/54 (2008): 321–330.

Balmer, Andrew S., Katie Bulpin, and Susan Molyneux-Hodgson. *Synthetic Biology: A Sociology of Changing Practices.* New York: Palgrave Macmillan, 2016.

Barthes, Roland. *The Rustle of Language.* Berkeley: University of California Press, 1989.

Bellow, Saul. *Herzog.* New York: Penguin, 1992.

Benjamin, Walter. "Surrealism: The Last Snapshot of the European Intelligentsia." *Se-lected Writings.* Vol. 2, Part 1, *1927–1930*, edited by Michael W. Jennings, Howard Ei-land, and Gary Smith, 206–221. Cambridge, MA: Belknap Press of Harvard University Press, 1992.

———. "Experience and Poverty." *Selected Writings.* Vol. 2, Part 2, *1931–1934*, edited by Michael W. Jennings, Howard Eiland, and Gary Smith, 731–738. Cambridge, MA: Belknap Press of Harvard University Press, 1999.

———. *The Arcades Project.* Translated by Howard Eiland and Kevin McLaughlin. Cambridge, MA: Harvard University Press, 2002.

Bennett, Gaymon. "What Is a Part? Biofab Human Practices Report 1.0." 2016. http://
biofab.synberc.org/sites/default/files/HPIP_DraftReport_Parts_1.0.pdf.

Biagioli, Mario, and Peter Galison, eds. *Scientific Authorship: Credit and Intellectual Property in Science*. New York: Routledge, 2003.

Blümel, Clemens. "Enrolling the Toggle Switch: Visionary Claims and the Capability of Modeling Objects in the Disciplinary Formation of Synthetic Biology." *NanoEthics* 10, no. 3 (2016): 269–287.

Boyer, Dominic. "Visiting Knowledge in Anthropology: An Introduction." *Ethnos* 70, no. 2 (2005): 141–148.

Brenner, Sydney. "Moron Peer Review." *Current Biology* 9, no. 2 (1999): 1–9.

Buesco, Yensi Flores, and Mark Tangney. "Synthetic Biology in the Driving Seat of the Bioeconomy." *Trends in Biotechnology* 35, no. 5 (2017): 373–378.

Bunzl, Matti. "The Quest for Anthropological Relevance: Borgesian Maps and Epistemological Pitfalls." *American Anthropologist* 110 (2008): 53–60.

Burnett, D. Graham. *Trying Leviathan: The Nineteenth-Century New York Court Case That Put the Whale on Trial and Challenged the Order of Nature*. Princeton, NJ: Princeton University Press, 2007.

———. "On the New Materialisms." *October* 155 (2016): 18–20.

Burnett, D. Graham, Matthew Rickard, and Jessica Terekhov, eds. *Keywords: For Further Consideration and Particularly Relevant to Academic Life*. Princeton, NJ: Princeton University Press, 2018.

Butkus, Ben. "Lilly and Princeton File Third IP-Infringement Suit against Generic Makers for Alimta Patent." GenomeWeb. May 6, 2009. https://www.genomeweb.com/biotechtransferweek/lilly-and-princeton-file-third-ip-infringement-suit-against-generic-makers-alimt#.WseNoWaZMb0.

Cabot, Heath. "The Business of Anthropology and the European Refugee Crisis." *American Ethnologist* 46, no. 3 (2019): 261–275.

Caduff, Carlo. "Speed Crash Course." *Cultural Anthropology* 32, no. 1 (2017): 12–20.

Calvert, Jane. "Ownership and Sharing in Synthetic Biology: A 'Diverse Ecology' of the Open and the Proprietary?" *BioSocieties* 7, no. 2 (2012): 169–187.

Campos, Luis. "That Was the Synthetic Biology That Was." In *Synthetic Biology: The Technoscience and Its Societal Consequence*, edited by Markus Schmidt, Alexander Kelle, Agomoni Ganguli-Mitra, and Huib Vriend, 5–21. Dordrecht: Springer, 2010.

Candea, Matei. "Arbitrary Locations: In Defense of the Bounded Field-Site." In *Multi-Sited Ethnography: Theory, Praxis and Locality in Contemporary Research*, edited by Mark-Anthony Falzon, 25–45. New York: Ashgate, 2009.

Carrier, Martin. "'Knowledge Is Power,' or How to Capture the Relationship between Science and Technoscience." In *Science Transformed? Debating Claims of an Epochal Break*, edited by Alfred Nordmann, Hans Radder, and Gregor Schiemann, 43–53. Pittsburgh, PA: Pittsburgh University Press, 2011.

Church, George, Michael Elowitz, Christina Smolke, Christopher Voigt, and Ron Weiss. "Realizing the Potential of Synthetic Biology." *Nature Reviews Molecular Cell Biology* 15, no. 4 (2014): 289.

Collins, Harry. *Changing Order: Replication and Induction in Scientific Practice*. Chicago: University of Chicago Press, 1992 [1985].

———. *Tacit and Explicit Knowledge.* Chicago: University of Chicago Press, 2010.

Collins, Harry, Robert Evans, and Martin Weinel. "STS as Science or Politics." *Social Studies of Science* 47, no. 4 (2017): 580–586.

Collins, James. "Bring in the Biologists." *Nature* 509, no. 7499 (2014): 155–156.

Collins, James Jr. "The Design Process for the Human Workplace." In *The Architecture of Science*, edited by Peter Galison and Emily Thompson, 399–412. Cambridge, MA: MIT Press, 1999.

Cook, Karen S., ed. *Trust in Society.* New York: Russell Sage, 2001.

Cooper, Melinda. *Life as Surplus: Biotechnology and Capitalism in the Neoliberal Era.* Seattle: University of Washington Press, 2008.

Creager, Angela N. H., Elizabeth Lunbeck, and M. Norton Wise. *Science without Laws: Model Systems, Cases, Exemplary Narratives.* Durham, NC: Duke University Press, 2007.

Csiszar, Alex. "Peer Review: Troubled from the Start." *Nature* 532 (2015): 306–308.

———. *The Scientific Journal: Authorship and the Politics of Knowledge in the Nineteenth Century.* Chicago: University of Chicago Press, 2018.

Dan-Cohen, Talia. "Epistemic Artefacts: On the Uses of Complexity in Anthropology." *Journal of the Royal Anthropological Institute* 23, no. 2 (2017): 285–301.

———. "Ignoring Complexity: Epistemic Wagers and Knowledge Practices among Synthetic Biologists." *Science, Technology, & Human Values* 41, no. 5 (2016): 899–921.

———. "Synthetic Biology in High Gloss." *BioSocieties* 13 (2018): 664–667.

Das, Veena, and Deborah Poole. "State and Its Margins: Comparative Ethnographies." In *Anthropology in the Margins of the State*, edited by Veena Das and Deborah Poole, 3–34. Santa Fe, NM: School of American Research Press, 1991.

Daston, Lorraine. "Objectivity and the Escape from Perspective." *Social Studies of Science* 22, no. 4 (1992): 597–618.

———. "The Moral Economy of Science." *Osiris* 10 (1995): 2–24.

———. "On Scientific Observation." *Isis* 99 (2008): 97–110.

Dilley, Roy, and Thomas G. Kirsch, eds. *Regimes of Ignorance: Anthropological Perspectives on the Production and Reproduction of Non-Knowledge.* New York: Berghahn Books, 2015.

Dizikes, Peter. "The Unraveling." *New York Times*, November 11, 2007.

Doing, Park. "Review Essay: Tacit Knowledge: Discovery by or Topic for Science Studies?" *Social Studies of Science* 41, no. 2 (2011): 301–306.

Edemariam, Aida. "Who's Afraid of the Campus Novel?" *The Guardian*, October 1, 2004.

Edgerton, David. "'The Linear Model' Did Not Exist: Reflections on the History and Historiography of Science and Research in Industry in the Twentieth Century." In *The Science–Industry Nexus: History, Policy, Implications*, edited by Karl Grandin, Nina Wormbs, and Sven Widmalm, 31–57. Sagamore Beach, MA: Science History Publications, 2004.

Elowitz, Michael B., and Stanislas Leibler. "A Synthetic Oscillatory Network of Transcriptional Regulators." *Nature* 403, no. 6767 (2000): 335–338.

Faulkner, Paul, and Thomas Simpson, eds. *The Philosophy of Trust.* Oxford: Oxford University Press, 2017.

Felski, Rita. "Critique and the Hermeneutics of Suspicion." *M/C Journal* 15, no. 1 (2012).

Feyerabend, Paul. *Against Method.* New York: Verso, 2002 [1975].

Finlay, Susan. "Engineering Biology? Exploring Rhetoric, Practice, Constraints and Collaborations within a Synthetic Biology Research Centre." *Engineering Studies* 5, no. 1 (2013): 26–41.

Fischer, Michael M. J. *Emergent Forms of Life and the Anthropological Voice.* Durham, NC: Duke University Press, 2003.

Fleck, Ludwik. *Genesis and Development of a Scientific Fact.* Chicago: University of Chicago Press, 1979.

Forman, Paul. "The Primacy of Science in Modernity, of Technology in Postmodernity, and of Ideology in the History of Technology." *History and Technology* 23, no. 1/2 (2007): 1–152.

Fortun, Mike. *Promising Genomics: Iceland and deCODE Genetics in a World of Speculation.* Berkeley: University of California Press, 2008.

Foucault, Michel. *The Order of Things.* New York: Vintage Books, 1994.

Freud, Sigmund. "Instincts and Their Vicissitudes." *The Complete Psychological Works.* Vol. 14. London: Hogarth Press, 1968.

"Frick Chemical Laboratory." Princeton University. http://etcweb.princeton.edu /CampusWWW/Companion/frick_chemical_laboratory.html.

Fyfe, Eileen. "Peer Review: Not as Old as You Might Think." *Times Higher Education (THE),* June 25, 2015. https://www.timeshighereducation.com/features/peer-review -not-old-you-might-think.

Galison, Peter. *Image and Logic: A Material Culture of Microphysics.* Chicago: University of Chicago Press, 1997.

———. "Buildings and the Subject of Science." In *The Architecture of Science,* edited by Peter Galison and Emily Thompson, 1–28. Cambridge, MA: MIT Press, 1999.

———. "The Collective Author." In *Scientific Authorship: Credit and Intellectual Property in Science,* edited by Mario Biagioli and Peter Galison, 325–357. New York: Routledge, 2003.

Gardner, Timothy S., Charles R. Cantor, and James J. Collins. "Construction of a Genetic Toggle Switch in *Escherichia coli.*" *Nature* 403, no. 6767 (2000): 339–342.

Gardner, Timothy S., and Kristy Hawkins. "Synthetic Biology: Evolution or Revolution? A Co-Founder's Perspective." *Current Opinion in Chemical Biology* 17 (2013): 871–877.

Geertz, Clifford. "Thick Description: Toward an Interpretive Theory of Culture." In *The Interpretation of Cultures,* 3–32. New York: Basic Books, 1973.

———. "The Last Humanist." *London Review of Books,* September 26, 2002.

Geroulanos, Stefanos. *Transparency in Postwar France: A Critical History of the Present.* Palo Alto, CA: Stanford University Press, 2017.

Giddens, Anthony. *The Consequences of Modernity.* Cambridge, UK: Polity Press, 2015.

Gieryn, Thomas F. "Two Faces on Science: Building Identities for Molecular Biology and Biotechnology." In *The Architecture of Science,* edited by Peter Galison and Emily Thompson, 423–457. Cambridge, MA: MIT Press, 1999.

Ginsberg, Alexandra Daisy, Jane Calvert, Pablo Schyfter, Listair Elfick, and Drew Endy. *Synthetic Aesthetics: Investigating Synthetic Biology's Designs on Nature.* Cambridge, MA: MIT Press, 2014.

Giordano, Sarah. "New Democratic Sciences, Ethics, and Proper Publics." *Science, Technology & Human Values* 43, no. 3 (2018): 401–430.

Gombrich, E. H. *Art and Illusion: A Study in the Psychology of Pictorial Representation.* Princeton, NJ: Princeton University Press, 2000 [1960].

———. *The Sense of Order: A Study in the Psychology of the Decorative Arts.* London: Phaidon Press, 2012.

Goodman, Nelson. *Languages of Art: An Approach to a Theory of Symbols.* Indianapolis, IN: Hackett, 1976.

———. *Ways of Worldmaking.* Cambridge, UK: Cambridge University Press, 1978.

Grasswick, Heidi E. "Scientific and Lay Communities: Earning Epistemic Trust through Knowledge Sharing." *Synthese* 177, no. 3 (2010): 387–409.

Green, Jeremy B. A., and James Sharpe, "Positional Information and Reaction-Diffusion: Two Big Ideas in Developmental Biology Combine." *Development* 142 (2015): 1203–1211.

Gross, Matthias. *Ignorance and Surprise: Science, Society, and Ecological Design.* Cambridge, MA: MIT Press, 2010.

Gusterson, Hugh. "Studying Up Revisited." *PoLAR* 20, no. 1 (2008): 114–119.

———. "Homework: Toward a Critical Ethnography of the University." *American Ethnologist* 44, no. 3 (2017): 435–450.

Hacking, Ian. *Representing and Intervening.* Cambridge, UK: Cambridge University Press, 1983.

———. "The Self-Vindication of the Laboratory Sciences." In *Science as Practice and Culture*, edited by Andrew Pickering, 29–64. Chicago: University of Chicago Press, 1992.

———. *The Social Construction of What?* Cambridge, MA: Harvard University Press, 2000.

Han, Byung-Chul. *The Transparency Society.* Palo Alto, CA: Stanford Briefs/Stanford University Press, 2015.

Hanson, Norwood Russell. *Patterns of Discovery: An Inquiry into the Conceptual Foundations of Science.* Cambridge, UK: Cambridge University Press, 1972.

Haraway, Donna J. *Primate Visions: Gender, Race, and Nature in the World of Modern Science.* London: Routledge, 1989.

———. *Modest_Witness@Second_Millenium.FemaleMan©_Meets_OncoMouse™: Feminism and Technoscience.* New York: Routledge, 1997.

Hardin, Russell. "Trustworthiness." *Ethics* 107, no. 1 (1996): 26–42.

———. *Trust and Trustworthiness.* New York: Russell Sage, 2004.

Harman, Oren, and Michael Dietrich. *Outsider Scientists: Routes to Innovation in Biology.* Chicago: University of Chicago Press, 2013.

Hayden, Erika Check. "Human Genome at Ten: Life Is Complicated." *Nature* 464, no. 7 (2010): 664–667.

Hedgecoe, Adam M. "Trust and Regulatory Organisations: The Role of Local Knowledge and Facework in Research Ethics Review." *Social Studies of Science* 42, no. 5 (2012): 662–683.

Helmreich, Stefan. *Alien Ocean: Anthropological Voyages in Microbial Seas.* Berkeley: University of California Press, 2009.

———. "What Was Life? Answers from Three Limit Biologies." *Critical Inquiry* 37, no. 4 (2011): 671–696.

——. *Sounding the Limits of Life: Essays in the Anthropology of Biology and Beyond.* Princeton, NJ: Princeton University Press, 2016.

Herper, Matthew. "Hype in Genes." *Forbes*, July 23, 2007. https://www.forbes.com/free _forbes/2007/0723/040.html.

Hilgartner, Stephen. "Adventurer in Genome Science." *Science* 318, no. 5854 (2007): 1244–1245.

Hirschauer, Stefan. "Editorial Judgments: A Praxeology of 'Voting' in Peer Review." *Social Studies of Science* 40, no. 1 (2010): 71–103.

ICTStalks. "The Architecture of Biological Complexity—Sydney Brenner." YouTube. January 24, 2013. https://www.youtube.com/watch?v=5uV73Vr0Os0.

Ingold, Tim. "Building, Dwelling, Living: How Animals and People Make Themselves at Home in the World." In *The Perception of the Environment: Essays on Livelihood, Dwelling, and Skill*, 172–188. London: Routledge, 2000.

Ingold, Tim, and Elizabeth Hallam. "Creativity and Cultural Improvisation: An Introduction." In *Creativity and Cultural Improvisation*, edited by Elizabeth Hallam and Tim Ingold, 1–24. Oxford: Berg, 2007.

Jacob, Francois, and Jacques Monod. "On the Regulation of Gene Activity." *Cold Spring Harbor Symposia on Quantitative Biology* 26 (1961): 193–211.

Kamtekar, Satwik, Jarad M. Schiffer, Huayu Xiong, Jennifer M. Babik, and Michael Hecht. "Protein Design by Binary Patterning of Polar and Nonpolar Amino Acids." *Science* 262, no. 5140 (1993): 1680–1685.

Keller, Evelyn Fox. *Refiguring Life: Metaphors of Twentieth-Century Biology*. New York: Columbia University Press, 1995.

——. *The Century of the Gene.* Cambridge, MA: Harvard University Press, 2000.

——. "What Does Synthetic Biology Have to Do with Biology?" *BioSocieties*. 4, no. 2/3 (2009): 291–302.

——. "Knowing as Making, Making as Knowing: The Many Lives of Synthetic Biology." *Biological Theory* 4, no. 4 (2009): 333–339.

Kelty, Christopher M. "Outlaw, Hackers, Victorian Amateurs: Diagnosing Public Participation in the Life Sciences Today." *Journal of Science Communication* 9, no. 1 (2010): C03. https://jcom.sissa.it/archive/09/01/Jcom0901(2010)C01/Jcom0901(2010) C03.

Kennedy, Rick. *A History of Reasonableness: Testimony and Authority in the Art of Thinking.* Rochester, NY: University of Rochester Press, 2004.

Kenney, Martin. *Biotechnology: The University-Industrial Complex.* New Haven, CT: Yale University Press, 1986.

Knorr Cetina, Karin. "Tinkering toward Success: Prelude to a Theory of Scientific Practice." *Theory and Society* 8, no. 3 (1979): 347–376.

——. *The Manufacture of Knowledge: An Essay on the Constructivist and Contextual Nature of Science.* New York: Pergamon Press, 1981.

Kramer, Roderick, and Karen Cook, eds. *Trust and Distrust in Organizations: Dilemmas and Approaches.* New York: Russell Sage, 2004.

Kuhn, Thomas. "Energy Conservation as an Example of Simultaneous Discovery." In *Critical Problems in the History of Science*, edited by Marshall Clagett, 321–356. Madison: University of Wisconsin Press, 1959.

———. *The Structure of Scientific Revolutions.* Chicago: University of Chicago Press, 2012 [1962].

Lamont, Michele. *How Professors Think: Inside the Curious World of Academic Judgment.* Cambridge, MA: Harvard University Press, 2009.

Landecker, Hannah. *Culturing Life: How Cells Became Technologies.* Cambridge, MA: Harvard University Press, 2007.

Lander, Eric, and Robert Weinberg. "Genomics: Journey to the Center of Biology." *Science* 287, no. 5459 (2000): 1777.

Langlitz, Nicolas. *Neuropsychedelia: The Revival of Hallucinogen Research since the Decade of the Brain.* Berkeley: University of California Press, 2012.

Latour, Bruno. *Science in Action.* Cambridge, MA: Harvard University Press, 1987.

———. *We Have Never Been Modern.* Cambridge, MA: Harvard University Press, 1993.

Latour, Bruno, and Steve Woolgar. *Laboratory Life: The Construction of Scientific Facts.* Princeton, NJ: Princeton University Press, 1986.

Law, John, and Annemarie Mol. *Complexities: Social Studies of Knowledge Practices.* Durham, NC: Duke University Press, 2002.

Lazebnik, Yuri. "Can a Biologist Fix a Radio?—Or, What I Learned while Studying Apoptosis." *Cell* 2, no. 3 (2002): 179–182.

Leitch, Alexander. *A Princeton Companion.* Princeton, NJ: Princeton University Press, 1978.

Lévi-Strauss, Claude. *Tristes Tropique: An Anthropological Study of Primitive Societies in Brazil.* Translated by John Russell. New York: Atheneum, 1967.

"Life Redesigned: The Emergence of Synthetic Biology." YouTube video. Vanderbilt University. www.youtube.com/watch?v=fXdzHY7wJHQ.

Lock, Margaret, and Patricia Kaufert. "Menopause, Local Biologies, and Cultures of Aging." *American Journal of Human Biology* 13, no. 4 (2001): 494–504.

Lovejoy, Arthur. *The Great Chain of Being.* Cambridge, MA: Harvard University Press, 1936.

Luhmann, Niklas. *Observations on Modernity.* Palo Alto, CA: Stanford University Press, 1998.

———. *Trust and Power.* Cambridge, UK: Polity Press, 2017.

Madrigal, Alexis. "Scientists Build First Man-Made Genome; Synthetic Life Comes Next." *Wired*, January 24, 2008.

———. "Synthetic Biology: It's Not What You Learned but What You Made." *Wired*, January 25, 2008.

———. "Wired Science Reveals Secret of Codes in Craig Venter's Artificial Genome." *Wired*, January 28, 2008.

Marcus, George E. *Ethnography through Thick and Thin.* Princeton, NJ: Princeton University Press, 1998.

Martin, Emily. "The End of the Body?" *American Ethnologist* 19, no. 1 (1992): 121–140.

Martin, Vincent, Douglas Pitera, Sydnor Withers, Jack Newman, and Jay Keasling. "Engineering a Mevalonate Pathway in Escherichia coli for Production of Terpenoids." *Nature Biotechnology* 21, no. 7 (2003): 796–802.

McGoey, Linsey. "The Logic of Strategic Ignorance." *British Journal of Sociology* 63, no. 3 (2012): 533–576.

Mead, Margaret. *Coming of Age in Samoa.* New York: Perennial Classics, 2001 [1928].

"Meet Princeton." Princeton University. https://www.princeton.edu/meet-princeton.

Merton, Robert. "The Self-Fulfilling Prophecy." *Antioch Review* 8, no. 2 (1948): 193–210.

Messeri, Lisa R. "The Problem with Pluto: Conflicting Cosmologies and the Classification of Planets." *Social Studies of Science* 40, no. 2 (2009): 187–214.

Meyer, Robinson. "How Gothic Architecture Took Over the American College Campus." *The Atlantic,* September 11, 2013. https://www.theatlantic.com/education/archive/2013/09/how-gothic-architecture-took-over-the-american-college-campus/279287/.

Mitchell, Robert. *Experimental Life: Vitalism in Romantic Science and Literature.* Baltimore, MD: Johns Hopkins University Press, 2013.

Mitchell, Sandra. *Unsimple Truths: Science, Complexity, and Policy.* Chicago: University of Chicago Press, 2009.

Miyazaki, Hirokazu, and Annelise Riles. "Teleonomic Mechanisms in Cellular Metabolism, Growth, and Differentiation." *Cold Spring Harbor Symposia on Quantitative Biology* 26 (1961): 389–401.

———. "Failure as an Endpoint." In *Global Assemblages: Technology, Politics, and Ethics as Anthropological Problems,* edited by Aihwa Ong and Stephen J. Collier, 320–333. Malden, MA: Blackwell, 2005.

Myers, Natasha. *Rendering Life Molecular: Models, Modelers, and Excitable Matter.* Durham, NC: Duke University Press, 2015.

Nader, Laura. "Up the Anthropologist—Perspectives Gained from Studying Up." In *Reinventing Anthropology,* edited by Dell Hymes, 284–311. New York: Pantheon Books, 1972.

"Naïve." *Oxford Dictionaries.* https://en.oxforddictionaries.com/definition/naive.

Nordmann, Alfred. "The Age of Technoscience." In *Science Transformed? Debating Claims of an Epochal Break,* edited by Alfred Nordmann, Hans Radder, and Gregor Schiemann, 19–30. Pittsburgh, PA: Pittsburgh University Press, 2011.

———. "Synthetic Biology at the Limits of Science." In *Synthetic Biology: Character and Impact,* edited by B. Giese, A. Von Gleich, C. Pade, and H. Wigger, 31–58. Berlin: Springer, 2014.

O'Malley, Maureen A. "Making Knowledge in Synthetic Biology: Design Meets Kludge." *Biological Theory* 4, no. 4 (2009): 378–389.

Panofsky, Erwin. *Gothic Architecture and Scholasticism.* New York: Meridian, 1957.

Peccoud, Jean, Megan Blauvelt, Yizhi Cai, Kristal Cooper, Oswald Crasta, Emily DeLalla, and Clive Evans. "Targeted Development of Registries of Biological Parts." *PLoS* 3, no. 7 (2008).

Pennisi, Elizabeth. "Synthetic Genome Brings New Life to Bacterium." *Science* 328 (2010): 958.

Pickering, Andrew. "Beyond Design: Cybernetics, Biological Computers, and Hylozoism." *Synthese* 168 (2009): 469–491.

Polanyi, Michael. *Personal Knowledge: Towards a Post-Critical Philosophy.* London: Routledge and Kegan Paul, 1958.

———. *The Tacit Dimension.* Chicago: University of Chicago Press, 2009.

Princess Bride, The. Directed by Rob Reiner. Culver City, CA: Act III Productions, 1987.

Proctor, Robert, and Londa L. Schiebinger, eds. *Agnotology: The Making and Unmaking of Ignorance*. Palo Alto, CA: Stanford University Press, 2008.

Quine, W. V. *Ontological Relativity and Other Essays*. New York: Columbia University Press, 1967.

Rabinow, Paul. *Essays in the Anthropology of Reason*. Princeton, NJ: Princeton University Press, 1996.

———. *The Accompaniment: Assembling the Contemporary*. Chicago: University of Chicago Press, 2011.

Rabinow, Paul, and Gaymon Bennett. *Designing Human Practices*. Chicago: University of Chicago Press, 2012.

Rabinow, Paul, and Talia Dan-Cohen. *A Machine to Make a Future: Biotech Chronicles*. Princeton, NJ: Princeton University Press, 2005.

Rabinow, Paul, and Anthony Stavrianakis. *Demands of the Day: On the Logic of Anthropological Inquiry*. Chicago: University of Chicago Press, 2013.

Raimbault, Benjamin, Jean-Philippe Cointet, and Pierre-Benoît Joly. "Mapping the Emergence of Synthetic Biology" *PLoS* 11, no. 9 (2016): 1–19.

Rasmussen, Nicolas. "Facts, Artifacts, and Mesosomes: Practicing Epistemology with the Electron Microscope." *Studies in the History and Philosophy of Science* 24, no. 2 (1993): 227–265.

Rees, Tobias. *After Ethnos*. Durham, NC: Duke University Press, 2018.

Rheinberger, Hans-Jörg. "Experimental Complexity in Biology: Some Epistemological and Historical Remarks." *Philosophy of Science* 64, no. 4 (1997): 245–254.

———. *Toward a History of Epistemic Things: Synthesizing Proteins in the Test Tube*. Stanford, CA: Stanford University Press, 1997.

———. "'Discourses of Circumstance': A Note on the Author in Science." In *Scientific Authorship: Credit and Intellectual Property in Science*, edited by Mario Biagioli and Peter Galison, 309–324. New York: Routledge, 2003.

———. "Gene Concepts: Fragments from the Perspective of Molecular Biology." In *The Concept of the Gene in Development and Evolution: Historical and Epistemological Perspectives*, edited by Peter J. Beurton, Raphael Falk, and Hans-Jörg Rheinberger, 219–239. Cambridge, UK: Cambridge University Press, 2008.

———. *An Epistemology of the Concrete: Twentieth-Century Histories of Life*. Durham, NC: Duke University Press, 2010.

Rheinberger, Hans-Jörg, and Jean-Paul Gaudillière. "Introduction." In *Classical Genetic Research and Its Legacy: The Mapping Cultures of Twentieth-Century Genetics*, edited by Hans-Jörg Rheinberger and Jean-Paul Gaudillière, 1–5. London: Routledge, 2006.

Ricoeur, Paul. *Freud and Philosophy: An Essay on Interpretation*. Translated by Denise Savage. New Haven, CT: Yale University Press, 1970.

Riles, Annelise. *The Network Inside Out*. Ann Arbor: University of Michigan Press, 2001.

Roosth, Sophia. "Biobricks and Crocheted Coral: Dispatches from the Life Sciences in the Age of Fabrication." *Science in Context* 26, no. 1 (2013): 153–171.

———. *Synthetic: How Life Got Made*. Chicago: University of Chicago Press, 2017.

Rosenfeld, Stuart, and Nalini Bhushan. "Chemical Synthesis: Complexity, Similarity, Natural Kinds, and the Evolution of a 'Logic.'" In *Of Minds and Molecules: New*

Philosophical Perspectives on Chemistry, edited by Nalini Bhushan and Stuart Rosenfeld, 187–207. Oxford: Oxford University Press, 2000.

Roth, Wolff-Michael. "Making Classifications (at) Work: Ordering Practices in Science." *Social Studies of Science* 35, no. 4 (2005): 581–621.

Roth, Wolff-Michael, and G. Michael Bowen. "'Creative Solutions' and 'Fibbing Results': Enculturation in Field Ecology." *Social Studies of Science* 31, no. 4 (2001): 533–536.

Ryle, Gilbert. *Collected Papers*, vol. 2. New York: Routledge, 1971.

———. *On Thinking*. Edited by Konstantin Kolenda. Oxford: Basil Blackwell, 1979.

Schyfter, Pablo. "How a 'Drive to Make' Shapes Synthetic Biology." *Studies in History and Philosophy of Biological and Biomedical Sciences* 44, no. 4 (2013): 632–640.

Sciszar, Alex. "Peer Review: Troubled from the Start." *Nature* 532 (2015).

———. *The Scientific Journal: Authorship and the Politics of Knowledge in the Nineteenth Century*. Chicago: University of Chicago Press, 2018.

Scott, James C. *Seeing Like a State: How Certain Schemes to Improve the Human Condition Have Failed*. New Haven, CT: Yale University Press, 1998.

Seasonwein, Johanna G. *Princeton and the Gothic Revival, 1870–1930*. Princeton, NJ: Princeton University Art Museum/Princeton University Press, 2012.

Shapin, Steven. *A Social History of Truth: Civility and Science in Seventeenth-Century England*. Chicago: University of Chicago Press, 1994.

———. *The Scientific Life: A Moral History of a Late Modern Vocation*. Chicago: University of Chicago Press, 2008.

———. *Never Pure: Historical Studies of Science as If It Was Produced by People with Bodies, Situated in Time, Space, Culture, and Society, and Struggling for Credibility and Authority*. Baltimore, MD: Johns Hopkins University Press, 2010.

Shapin, Steven, and Simon Schaffer. *Leviathan and the Air-Pump: Hobbes, Boyle, and the Experimental Life*. Princeton, NJ: Princeton University Press, 2011.

Shavit, Ayelet, and Aaron M. Ellison. "Toward a Taxonomy of Scientific Replication." In *Stepping in the Same River Twice: Replication in Biological Research*, edited by Ayelet Shavit and Aaron M. Ellison. 22nd ed. Vol. 3. New Haven, CT: Yale University Press, 2017.

Simmel, Georg. *The Sociology of Georg Simmel*. Edited and translated by K. H. Wolff. New York: Free Press, 1950.

Simpson, Michael, Chris D. Cox, Gregory D. Peterson, and Gary S. Sayler. "Engineering in the Biological Substrate: Information Processing in Genetic Circuits." *Proceedings of the IEEE* 92, no. 5 (2004): 846–862.

Sismondo, Sergio. "Post-Truth?" *Social Studies of Science* 47, no. 1 (2017): 3–6.

Skinner, Quentin. "Meaning and Understanding in the History of Ideas." *History and Theory* 8, no. 1 (1969): 3–53.

Smith, Richard. "Peer Review: A Flawed Process at the Heart of Science and Journals." *Journal of the Royal Society of Medicine* 99, no. 4 (2006): 178.

Smithson, Michael. *Ignorance and Uncertainty: Emerging Paradigms*. New York: Springer, 1989.

Star, Susan, and James Griesemer. "Institutional Ecology, 'Translations,' and Boundary Objects: Amateurs and Professionals in Berkeley's Museum of Vertebrate Zoology, 1907–39." *Social Studies of Science* 19, no. 3 (1989): 387–420.

Stevens, Ruth. "Elements of New Frick Lab Join to Create 'Best Infrastructure' for Chemistry." Princeton University. September 2, 2010. https://www.princeton.edu/news /2010/09/02/elements-new-frick-lab-join-create-best-infrastructure-chemistry.

Strathern, Marilyn. "Cutting the Network." *Journal of the Royal Anthropological Institute* 2, no. 3 (1996): 517–535.

———. "The Tyranny of Transparency." *British Educational Research Journal* 26, no. 3 (2000): 309–321.

Sunder Rajan, Kaushik. *Biocapital: The Constitution of Postgenomic Life.* Durham, NC: Duke University Press, 2006.

Thomas, Kedron. *Regulating Style: Intellectual Property Law and the Business of Fashion in Guatemala.* Berkeley: University of California Press, 2016.

Traweek, Sharon. *Beamtimes and Lifetimes: The World of High Energy Physics.* Cambridge, MA: Harvard University Press, 1988.

Turing, Alan M. "The Chemical Basis of Morphogenesis." *Philosophical Transactions of the Royal Society B* 237, no. 641 (1952): 5–72.

Vanderbilt University. "Life Redesigned: The Emergence of Synthetic Biology." YouTube video. October 25, 2013. https://www.youtube.com/watch?v=fXdzHY7wJHQ.

Venter, J. Craig, and Daniel Cohen. "The Century of Biology." *NPQ New Perspectives Quarterly* 21, no. 4 (2004): 73–77.

Vidler, Anthony. *The Architectural Uncanny: Essays in the Modern Unhomely.* Cambridge, MA: MIT Press, 1992.

Voosen, Paul. "Synthetic Biology Comes Down to Earth." *Chronicle of Higher Education.* March 4, 2013. https://www.chronicle.com/article/Synthetic-Biology-Comes -Down/137587.

Wade, Nicholas. "Scientists Transplant Genome of Bacteria." *New York Times,* June 29, 2007.

———. "Genetic Engineers Who Don't Just Tinker." *New York Times,* July 8, 2007.

Way, Jeffrey C., James Collins, Jay Keasling, and Pamela Silver. "Integrating Biological Redesign: Where Synthetic Biology Came From and Where It Needs to Go." *Cell* 157, no. 1 (2014): 151–161.

Weber, Marcel. *Philosophy of Experimental Biology.* Cambridge, UK: Cambridge University Press, 2005.

Weber, Max. "Science as a Vocation." In *From Max Weber: Essays in Sociology,* translated and edited by H. H. Gerth and C. Wright Mills, 129–156. New York: Oxford University Press, 1946.

Whyte, Kyle Powys, and Robert P. Crease. "Trust, Expertise, and the Philosophy of Science." *Synthese* 177, no. 3 (2010): 411–425.

Wood, Christopher S. "Art History Reviewed VI: E. H. Gombrich's 'Art and Illusion: A Study in the Psychology of Pictorial Representation.'" *Burlington Magazine* (2009): 836–839.

Zandonella, Catherine. "Alimta: Fundamental Science, Fundamental Benefit." Princeton University. November 18, 2011. http://research.princeton.edu/news/features/a /index.xml?id=6187.

Zuckerman, Harriet, and Robert K. Merton. "Patterns of Evaluation in Science: Institutionalization, Structure, and Functions of Referee Systems." *Minerva* 9 (1971): 66–100.

INDEX

CPSIA information can be obtained
at www.ICGtesting.com
Printed in the USA
LVHW031004120221
679143LV00001B/104

9 781501 754333